From Farm to Fork

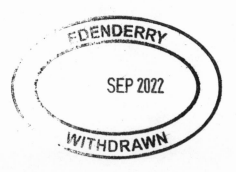

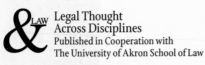

From Farm to Fork

Perspectives on Growing Sustainable Food Systems in the Twenty-First Century

Edited by Sarah J. Morath

University of Akron Press
Akron, Ohio

20   19   18   17   16   5   4   3   2   1

ISBN: 978-1-629220-10-9 (paper)
ISBN: 978-1-629220-11-6 (ePDF)
ISBN: 978-1-629220-12-3 (ePub)

LIBRARY OF CONGRESS CATALOGING-IN-PUBLICATION DATA
Names: Morath, Sarah, editor.
Title: From farm to fork : perspectives on growing sustainable food systems in the twenty-first century / Sarah Morath, editor.
Description: First edition. | Akron, Ohio : University of Akron Press, [2016] | Series: &law | Includes bibliographical references and index.
Identifiers: LCCN 2016025856 (print) | LCCN 2016030949 (ebook) | ISBN 9781629220109 (pbk. : alk. paper) | ISBN 9781629220116 (ePDF) | ISBN 9781629220123 (ePub)
Subjects: LCSH: Food supply—Environmental aspects—United States. | Agriculture—Environmental aspects—United States. | Sustainable agriculture—United States.
Classification: LCC HD9005 .F756 2016 (print) | LCC HD9005 (ebook) | DDC 338.10973—DC23
LC record available at https://lccn.loc.gov/2016025856

∞ The paper used in this publication meets the minimum requirements of ANSI / NISO Z39.48–1992 (Permanence of Paper).

Cover design: Amy Freels. Photo by Amy Freels, copyright © 2014. Used with permission.

From Farm to Fork was designed and typeset by Amy Freels, with assistance from Tyler Krusinski. The typeface, Stone Print, was designed by Sumner Stone in 1991. *From Farm to Fork* was printed on sixty-pound natural and bound by Bookmasters of Ashland, Ohio.

Jill K. Clark, Shoshanah Inwood, and Jeff S. Sharp, *The Social Sustainability of Family Farms in Local Food Systems: Issues and Policy Questions*. Reprinted by permission of the Publishers from *Local Food Systems: The Birth of New Farmers and the Demise of the Family Farm?*, in Local Food Systems in Old Industrial Regions eds. Neil Reid, Jay D. Gatrell, and Paula S. Ross (Farnham: Ashgate, 2012), pp. 131–145. Copyright © 2012.

Jason J. Czarnezki, *Informational and Structural Changes for a Sustainable Food System*. An earlier version was published in 31 Utah Envtl. L. Rev. 263 (2011).

Marion Nestle, *Utopian Dream: A Farm Bill Linking Agriculture to Health*. Originally appeared as Marion Nestle, *Utopian Dream: A New Farm Bill*, in Dissent 2012, 15–19. Reprinted with permission of the University of Pennsylvania Press.

Susan A. Schneider, *A Call for the Law of Food, Farming, and Sustainability*. Parts of this article are drawn from A Reconsideration of Agricultural Law: A Call for the Law of Food, Farming, and Sustainability, 34 Wm. & Mary Envtl. L. & Pol'y Rev. 935 (2010) and Food Farming & Sustainability: Readings in Agricultural Law (2011).

Contents

Contributors

Molly D. Anderson, William R. Kenan Jr. Professor of Food Studies, Middlebury College. B.S., M.S. Colorado State University; Ph.D. University of North Carolina at Chapel Hill (Systems Ecology).

Mary Jane Angelo, Professor of Law, Director of Environmental and Law Use Program, University of Florida. B.S. Rutgers University; M.S. University of Florida; J.D. University of Florida.

Jill K. Clark, Assistant Professor, John Glenn School of Public Affairs, Ohio State University. B.S. Ohio State University; M.S. University of Wisconsin; Ph.D. Ohio State University (Geography).

Jason J. Czarnezki, Gilbert and Sarah Kerlin Distinguished Professor of Environmental Law, Elisabeth Haub School of Law at Pace University. B.A. University of Chicago; J.D. University of Chicago.

Oran B. Hesterman, President and CEO, Fair Food Network. B.S., M.S. University of California–Davis; Ph.D. University of Minnesota (Agronomy, Plant Genetics, and Business Administration).

John Ikerd, Professor Emeritus of Agricultural and Applied Economics, University of Missouri Columbia, College of Agriculture, Food and Natural Resources. B.S., M.S., Ph.D. University of Missouri (Agricultural Economics).

Shoshanah Inwood, Assistant Professor, Community Development and Applied Economics, University of Vermont. B.A. Oberlin College, M.S., Ph.D. Ohio State University (Rural Sociology).

Saru Jayaraman, Director, Food Labor Research Center, University of California, Berkeley. B.A. University of California–Los Angeles; M.P.P. Harvard University; J.D. Yale Law School.

Jane Kolodinsky, Professor and Chair, Community Development and Applied Economics, University of Vermont. B.S., M.B.A. Kent State University; Ph.D. Cornell University (Consumer Economics).

Caitlin R. Marquis, Healthy Hampshire Coordinator, Collaborative for Educational Services. B.A. The George Washington University; M.S. The Ohio State University.

Sarah J. Morath, Clinical Associate Professor, University of Houston Law Center. B.A. Vassar College; M.E.S. Yale University; J.D. University of Montana School of Law.

Marion Nestle, Paulette Goddard Professor, Department of Nutrition, Food Studies, and Public Health, New York University. M.P.H., Ph.D. University of California, Berkeley (Molecular Biology).

Susan A. Schneider, Director of the LL.M. Program in Agricultural and Food Law, Professor of Law, University of Arkansas. B.A. College of St. Catherine; J.D. University of Minnesota; LL.M University of Arkansas (Agricultural Law).

Jeff S. Sharp, Director and Professor of Rural Sociology, College of Food, Agriculture, and Environmental Sciences, Ohio State University. B.A., M.S., Ph.D. Iowa State University (Sociology).

Josh Slotnick, PEAS (Program in Ecological Agriculture and Society) Director, University of Montana Professor, Clark Fork Organics Co-Founder. B.A. University of Montana; M.S. Cornell University.

Foreword

Oran B. Hesterman, Fair Food Network

My first exposure to the sustainable food system movement happened in the early 1970s, when I was a student at the University of California, Santa Cruz. As a twenty-year-old sophomore, I was attracted to the Farm, an innovative project located on seventeen acres of rich, fertile soil and inspired by the principles of biodynamic agriculture, with a clear view of the ever-changing Monterey Bay. It was here that I came to understand that the food system as it was then functioning would not sustain our global population, which is increasing at an alarming rate. And I was living, day to day, in a different relationship with the earth and farming in a different way that could, in fact, prove to be an alternative model. The Farm has since become the Center for Agroecology and Sustainable Food Systems, a training ground for hundreds of apprentices in organic farming techniques, and one of the many places where young people have been finding ways to fuel the movement, which has grown by leaps and bounds since those days more than forty years ago.

Since that time, the sustainable agriculture or "good food" movement has grown in many directions to include not only more ecologically sound farming, but also issues of social and racial equity, just and fair treatment of farm and food workers, equitable access to affordable healthy food, and public health consequences of a food system that produces too much of what is not healthy for our bodies and too little of what is.

The authors of the chapters in this book delve into these issues and others from a variety of perspectives and offer practice and policy solutions to put the food system back on track for our children, our communities, and our environment. Books like this one are important as blueprints for redesigning our food system. We have made great strides as a movement in the past four decades. At the same time, our policy experts point out in their chapters how much more we need to accomplish.

In reality, food connects us like few other things. It reflects our cultures, our traditions, and our rituals, and is our most profound and sensual connection to the earth. It nurtures and sustains us, heals our bodies, or, in its lack or excess, creates disease. Food touches everything. Climate and conservation are affected by what we grow and how we grow it. Food—production, processing, distribution, service—makes up between 4 and 8 percent of our national economy and accounts for close to 15 percent of all jobs.[1] It is the second largest economic sector in my home state of Michigan. To this day, it can spark revolutions, as we saw with food price instability and the Arab Spring uprisings in 2010–2011.

Yet when most of us think of food, we think about what is for dinner. We might be thinking about what is in our refrigerators. Those of us who are particularly interested might think about which grocery store to shop at, whether they carry locally grown or organic products, or whether the farmers' market would be a better choice this week.

The biggest challenge in terms of shifting the food system is that we are not thinking of food in terms of a system, but rather about how food affects us as individuals. But the who, what, when, where, and how of what we eat is broader than the individual. It is a system. And there are no systems of one.

Think about our system of streets, roads, and highways. If we were to look at this system the way we tend to look at food, we would expect each person to fix the potholes in front of his or her house. We understand that streets, roads, and highways are part of a whole transportation system: collectively, we take responsibility for it. We know that it cannot function successfully if we consider it only as it relates to each individual. Similarly, we cannot significantly shift our food system when everyone thinks about it individually.

We—and by we, I mean people, communities, governments, businesses, and nonprofits—all must understand and claim the potential for food to transform our lives, our towns, and our planet. This is not about realizing potential—this is a necessity. Food is a component of the biggest problems that face our country and planet: global warming, health care and national budget, population growth, and income inequality.

For example, the largest aquatic dead zone in the United States is in the Gulf of Mexico: 8,500 square miles of water with so little oxygen that the creatures that should live there have either died or fled. And that dead zone is caused in large part by agricultural runoff that flows into the Mississippi River from Minnesota to Louisiana.

I experienced this dead zone firsthand, shrimping on the Gulf with Ray Bradhurst, a third-generation shrimper from Louisiana. Hours before the break of dawn, we stepped onto his boat and started our journey. Yet it took us three hours of racing across a vast area in the Gulf just to get to water that would sustain life and from which Ray could harvest shrimp to make his living. It is hard for most of us to imagine traveling a distance the size of the state of Massachusetts (the current size of the dead zone) just to get to work—but for Ray and other shrimpers, this is the reality they face.

Our nation's health bills are skyrocketing in part because of preventable, diet-related illnesses such as diabetes and heart disease. And it starts with our youth. Though we have received some heartening news recently about obesity among the very young, the overall numbers remain disturbing. Seventeen percent of kids aged two to nineteen are obese. That number jumps three points for African American kids and five points for Latino youth. This is unconscionable. And it is not only affecting our children. Reflective of the progression of the disease, the rate of obesity for adults was 14.5 percent in 1974 and 36.1 percent in 2010.[2] At this rate of increase, this diet-related disease, with its concomitant complications, will dramatically undermine our health care system and bankrupt us as a nation.

Meanwhile, we have a national economic policy that continues to favor massive commodity crops over fresh, healthy food. But there is hope on that front as well. The number of family farms increased by 400,000 between 1996 and 2012.[3] And the number of farmers' markets in the United States has grown from 340 in the early 1970s to more than 8,000 in 2014, a remarkable indication of the growing awareness of individuals nationwide who understand the value of consuming locally grown food.

This is all truly great news. But if people see each of these as isolated successes to isolated problems, we will not amass the collective will we need to make the substantial changes required.

It is imperative that we shift our view of food to be comprehensive, integrated, and cohesive. That is a big transformation. You might think about it as something akin to how we see media today versus how we saw it twenty-five years ago. Back then, you cracked open a book. You unfolded your newspaper. You went to the cinema. You watched your favorite TV show at the same time every week along with the rest of the country. Now your phone or tablet is your library, TV set, movie house, and newsstand. We consume media when we want, how we want, and often in real time.

That is a tremendous change and the kind of shift in understanding we need to happen around food. All of us need to think of it not just as an ingredient on our plates. It is there, yes, but before that it was in the ground; then it was in someone's hands. Is that person paid adequately? Is that food available for your less wealthy neighbors as well? In all neighborhoods? Are our national policies and local regulations supportive of growing and processing it nearby?

Systems that truly work well as systems are undergirded by principles. Our highway system is based on the principles that roads should facilitate safe, speedy travel.

Our food system must also be defined by core principles. It should be healthy; it should be green—grown by environmentally sustainable methods; it should be affordable, so all have access to the food needed to lead a healthy life; and it should be fair, meaning those who produce it work under good labor practices and receive fair wages.

Healthy, green, fair, and affordable—these are the principles of a good food system.

So, the question is: how do we take the pockets of success the Good Food Movement has seen in the past three decades and scale them more broadly? How do we work to ensure that they are not isolated victories but are big wins that influence the system as a whole? How do we foster an understanding among our neighbors that food is not an ingredient or a meal but a system that affects our communities, country, and planet? And how do we build the political will so that food system policy change can happen at a faster and more transformational pace?

I think we can make significant progress in these directions by building three bridges. We need to build bridges between traditionally siloed food issues; we need to build bridges that take successful projects from model to mainstream; and we need to build bridges between the partners with whom we work.

The first bridge we need to build is between issues. Think of it as moving from *silos* to *systems*. We need more solutions that encompass multiple wins, which is what you hope happens when you look for systems solutions rather than trying to solve problems one at a time.

There are times when groups working on food issues seem to be camped at one of two poles. On one side are the epicureans or foodies, who implore society to get used to paying the "real cost" of good food; that is, if we want fresh produce and healthy meat—rather than processed corn and soy products—we need to pay more for the labor that goes into growing and harvesting them. Make no mistake: the thinking here is commendable.

At the other pole are antihunger activists, for whom the top priority is preserving calories for those who are most vulnerable. And of course, making sure families have enough to eat should be a concern for all of us. It is shameful that people go hungry in our country.

But both the epicureans and antihunger activists are held back by limits on their visions. If we focus, as the epicureans would have us do, solely on "real costs," then healthy food is in danger of becoming a luxury for the elite. And yet if we focus solely on the protection of calories, the opposite happens. We end up with national policy focused on keeping food *very* inexpensive, which ignores much-needed agricultural improvements and support for family farmers and for farm and food workers. There should not be a question of whether we support hungry families or local farmers—we can do both. We need solutions that deliver multiple wins.

In Philadelphia, a dynamic organization called Common Market is creating such a multiple-win solution. They are connecting wholesale food customers with farmers in the regions surrounding Philadelphia and marketing good food (healthy, green, fair, and affordable) to schools, hospitals, grocers, faith-based institutions, and workplaces. Common Market works with seventy-five regional farmers and aggregates their production. Farmers are offered a fair price for what they produce, and they offer Common Market's customers a year-round source of local food. Common Market also has a mission of serving vulnerable communities. Many of the students, patients, clients, and workers in the institutions that Common Market serves come from underserved communities. Common Market employs people from those communities and has also found ways to bring healthy food into public schools and groceries in low-income neighborhoods at prices that are affordable. Common Market is an example of an organization focusing on *systems* rather than *silos,* and we all have a lot to learn from this model.

The second bridge we need to build is from *model* to *mainstream*. We have many, many success stories that did not exist twenty years ago. Too often, however, these projects are isolated. They may be having a tremendous impact in a small area or for a few families, but they could be having much greater impact if they could be taken to scale (over a greater geographical area, involving a greater number of people, etc.). One reason for the lack of scale on the part of many community-based projects is that the people running those projects (whether a community garden, a mobile meat processor, or a healthy corner store initiative) are likely focused squarely on making a difference in their community. They might like others to adopt their model, but expanding to other states, for instance, is not a priority.

That is where public policy comes in. Policy can help us think about small successes in terms of systems and broader replication; it can be the vehicle that carries us from model to mainstream.

One model that has entered more mainstream practice is farm to school. We all should be proud of the progress that this aspect of the movement has made in a relatively short time. The first farm to school programs sprouted in California and Florida in 1996. Now there are farm to school programs in all 50 states, tapping into the deep interest of parents, schools, children, and local communities in providing fresh, local fruits and vegetables for our children.[4] Farm to school projects, which support school gardens, nutrition education, and local produce in cafeterias, have received federal policy support for several years, giving us an example of model to mainstream. Today, farm to school projects reach 21 million students in nearly 40,000 schools. That is almost 40 percent of the student population. This rapid adoption of farm to school programs resulted from smart tactical moves on the part of leaders around the country, and also reflects a cultural shift that we see at the highest levels of our federal government, with the First Lady's vocal priorities.

I have had an opportunity to see firsthand the power of public policy to help make the shift from model to mainstream at Fair Food Network. Our program, known as Double Up Food Bucks, is an example of a "healthy food incentive program." Throughout Michigan, people who use their Supplemental Nutrition Assistance Program benefits (SNAP, formerly called food stamps) at participating sites receive a one-to-one match to purchase healthy, locally grown fruit and vegetables, up to $20 at farmers' markets every market day. The wins are threefold: families bring home more healthy food; farmers gain new customers and make more money; and more food dollars stay in the local economy. Each has a ripple effect of benefits across the community. The project started at five farmers' markets in Detroit in 2009; six years later, it is at more than 150 sites across the state including grocery stores in one of the first pilots in the nation.

There are farmers in Michigan, such as Vicki Zilke, who say that half their sales are now from SNAP customers. Other farmers are telling us that Double Up Food Bucks has brought their vegetables to a new community of customers, saying: "I'm glad to have more business, but even aside from the sales factor, I'm happy knowing the people have the good food." And the SNAP participants agree. Our evaluations document that more than 85 percent of SNAP customers are buying and eating more fruits and vegetables when Double Up Food Bucks incentives are available.

Michigan is not the only state in which healthy food incentives are being implemented. Models similar to Double Up are now in hundreds of farmers' markets in communities throughout the country. Several organizations (Fair Food Network, Market Umbrella, Roots of Change, Wholesome Wave) banded together to conduct a cluster evaluation of these programs, with the intent to use the evaluation to inform nutrition policy in the Farm Bill. We knew that bringing the Double Up idea from model to mainstream required a shift in public policy. In the view of many of us in the Good Food Movement, the 2014 Farm Bill contained its share of disappointments, but there were bright spots. Having seen the positive impact of Double Up Food Bucks in her home state of Michigan, Senator Debbie Stabenow became a steadfast champion of including a provision in the Farm Bill to start scaling this idea. The Farm Bill signed into law in February 2014 included $100 million to expand programs like Double Up Food Bucks across the nation. This is a big win. With the number of Americans using food stamps in this country increasing from 2.8 million in 1969 to more than 45 million in 2014 (bringing it close to 15 percent of the U.S. population), more federal money will be available to support healthier eating habits for low-income consumers while supporting local economies.

While building bridges from silos to systems and model to mainstream will support a more sustainable food system, the big question remaining is "How?" What will it take to build these bridges and use them to transform our food system into one that is healthy, green, fair, and affordable?

I believe that it takes a third bridge to accomplish the first two; that is, the bridge from *them* to *us*. We must expand who the Good Food Movement works with. I know that it is often more comfortable to work with groups and leaders whose goals are in alignment with our own; however, if we are not more inclusive, we are not going to see the change we need in the time frame we need to make a significant impact on problems such as global warming, our health crisis, and income inequality. The more of us in the Good Food Movement who spend our energy identifying the enemy and entering into battles with them, the more we are diverted into unwinnable fights. We need to focus on finding and implementing workable ideas versus spending our time staking out ideological ground. I believe we are being called on to be food system "solutionaries."

We must engage partners at every level in designing a food system that works for "them" as well as "us." Who do I mean by them? It could be large-scale producers or big agricultural companies. It could be conventional distributors or restaurants. It could be politicians who sit across the aisle from our natural comfort zone. Simply put, "they" are people and organizations involved in food who may not yet be involved in the Good Food Movement.

Let's look at the Coalition of Immokalee Workers (CIW), which represents workers on tomato farms. Their Campaign for Fair Food has won huge victories for the treatment and pay of farmworkers in the Immokalee region of southwest Florida.

Initially, they were focused on fighting with the growers for better working conditions and fairer wages. They did not get very far—the growers were being squeezed between greater worker demands and demands for low prices from their customers. Then CIW took their solution to the biggest buyers of their product: Taco Bell, Burger King, Aramark, Whole Foods, Sodexo, Subway, and other restaurants and food suppliers. By securing agreements that these buyers pay an additional penny and a half per pound of tomatoes and that this additional money would go directly to the farmworkers, the workers increased their pay by more than 50 percent. Just as important, the Immokalee Workers have, as they like to say, transformed Florida's fields "into a workplace rooted in mutual respect and basic dignity for farmworkers."[5]

The Good Food Movement must expand its reach to have the impact we want and need. If potential partners are not yet aligned with our work, we cannot shun them. We need instead to bring them along and meet them on whatever common ground exists that can help us build a closer relationship. In my view, that is how you make changes to whole systems.

Not everyone is going to be an active participant in the Good Food Movement—but you do not need everybody! The civil rights movement did not need everyone in the country on its side to demolish Jim Crow. The LGBT community has made massive strides for marriage equality in the past decade, including the recent Supreme Court decision, without the entire country at its side.

What we need are *enough* people who understand that food is a system that affects not just our meals, but our communities, our country, and our planet. We can help create that critical mass by building bridges: from Silos to Systems, from Model to Mainstream, from Them to Us.

In the early 1970s there were only a few of us experimenting with organic farming and local food systems. It has now become a powerful cultural trend. I believe that many more people are ready to shift their awareness to see food as a system and to see themselves as change agents.

The authors of the chapters that follow, many of whom are seasoned and well-established leaders in the field, offer solid, tangible ideas about systems solutions and policy changes that can help move the Good Food Movement forward faster. These are ideas that we all need to pay attention to, follow with enthusiasm, and lend support to in every way we can. We can all see ourselves as bridge builders, and some of the plans for those bridges are written here.

NOTES

1. E-mail from Michael Shuman, Teaching Adjunct, Simon Fraser University (June 27, 2014).

2. Rich Pirog, Crystal Miller, Lindsay Way, Christina Hazekamp, and Emily Kim, *Good Food Timeline, The Local Food Movement: Setting the Stage for Good Food*, http://foodsystems .msu.edu/uploads/files/Good_Food_Timeline_WEB.pdf.

3. USDA Economic Research Service, *Family and Nonfamily Farms, by Farm Size Class (Gross Sales), 1996–2012*, http://www.ers.usda.gov/data-products/farm-household-income-and -characteristics.aspx#.U6EXFI1dV9k (follow "Family and Nonfamily Farms).

4. USDA Food and Nutrition Service, *The Farm to School Census*, http://www.fns.usda.gov /farmtoschool/census#/.

5. Coalition of Immokalee Workers, *Campaign for Fair Food*, http://ciw-online.org/campaign -for-fair-food/.

Preface

Sarah J. Morath, University of Houston Law Center

Interest in the food we eat and how it is produced, distributed, and consumed has grown tremendously in the last decade. In droves, people are exchanging highly processed, genetically engineered, chemically laden food for locally grown organic products. The growth of farmers' markets from 1,755 in 1994 to more than 8,200 in 2014, in both urban and rural areas, is just one indication that consumers are interested in knowing who grew their food and how that food was grown.[1] Increasingly, policy makers, academics, and community leaders are seeing the food we eat as part of a larger system—a food system involving environmental, economic, health, community, and worker concerns.

In all fairness, this book does not address every aspect of our complex food system. A single book could not. Instead, it is a starting point for a larger discussion on what is needed to create a sustainable food system. The book brings together experts in the fields of law, economics, nutrition, and social sciences, as well as farmers and advocates. These experts share their perspectives on and suggestions for creating healthy, sustainable, and equitable food systems in the future.

I'd like to thank all the contributors for sharing their experiences and expertise and for continuing to look for solutions. I'd also like to thank the editors at the University of Akron Press for their assistance and guidance. Finally, I would not have been able to complete this project without the help of my reliable and hardworking research assistant, Monica Dobson.

NOTES

1. USDA Economic Research Service, *Number of U.S. Farmers' Markets Continues to Rise* (2016), http://ers.usda.gov/data-products/chart-gallery/detail.aspx?chartId=48561.

Introduction

Sarah J. Morath, University of Houston Law Center

L
ike any system, our food system consists of a complex web of players and processes. Our food system is influenced by farmers, consumers, businesses, and policy makers who produce, harvest, distribute, prepare, and dispose. Creating a sustainable food system will require an integrated approach from a variety of disciplines, including nutrition, agroecology, law, economics, and consumer sciences. Given the intrinsic complexity of our food system, most books on the topic cannot address every aspect of our food system. This book is no exception. Rather than providing a singular explanation on creating a sustainable food system, this book provides the perspectives of numerous academics and advocates whose work focuses on different parts of our food chain. These perspectives appear in three parts. The first part describes a few of the elements that comprise our food system. The second part expresses the view of some of the players within the system. The third and final part proposes potential solutions to making our food system more sustainable.

Part I, The Elements of Our Complex Food System, begins with a critique of the piece of legislation that has the greatest influence on our food system: the farm bill. Marion Nestle notes that the current structure of the farm bill favors big agriculture over small organic farmers and encourages growing commodity crops, like corn and soy, over fruits and vegetables, a practice that Nestle argues encourages weight gain. Nestle's utopian farm bill would better support farmers, the environment, and human health.

In the following chapter, John Ikerk questions whether there will be enough land for farming in the twenty-first century and highlights the importance of creating a food-secure future. Ikerk argues that current industrial farming practices will not be able to meet the meet the food and nutrition needs of the future. As alternatives, Ikerk proposes "permanently zoning" land for food production and a society where everyone is given the assurance of an income adequate to meet his or her essential economic needs, including enough good food to support healthy, active lifestyles.

Ikerk's chapter is followed by a chapter by Jill Clark, Jeff Sharp, and Shoshona Inwood. These three have studied the intergenerational succession of family farms to assess the long-term viability and sustainability of farm businesses engaged in sustainable food systems. Clark, Sharp, and Inwood argue that farm succession must take into account an influx of a new generation of farmers, or first generation farmers (FGF), drawn to the sustainable farm movement. The three note differences between multigenerational farm families and first-generation farm families, and caution that such difference may influence the long-term persistence of working agricultural landscapes in exurban areas.

The first part concludes with a chapter by Molly Anderson, who describes how social sustainability, wages, and working conditions are being addressed in the US food system through voluntary standards and state and municipal food plans. She argues that food security will require a rights-based approach and democratic decision-making that includes the voices of vulnerable and marginalized people.

Part II, Views From Within the System, considers the perspectives of the farmer, consumer expert, and the food worker advocate. This section begins with a chapter by Josh Slotnik, who describes his experience farming in the Missoula Valley in Montana. He contrasts his experience in community agriculture, a term that he explains more accurately captures the effect of urban agriculture, to that of industrial agriculture, one that is segmented and placeless, the origins of our food, unknown. Slotnik argues that farming together through community agriculture is a way to cultivate fairness, justice, and an environmental ethic.

The perspective of the consumer is provided in the next chapter by Jane Kolondinsky. Kolondinsky describes where people access food, what food is available at these access points, and how consumer decisions can foster or inhibit sustainable food systems. Kolondinsky describes several purchasing points including gardens (home and community), community-supported agriculture (CSA), farmers' markets, community stores (general stores and independent grocery stores), supermarkets and superstores, and institutional purchasing. She argues that the alternative access points such as home gardens, CSAs, farmers' markets, and institutional purchasing offer the greatest opportunities for creating a sustainable food system.

Saru Jayaraman concludes this part with a chapter on an element of the food system often overlooked: the retail and restaurant worker. Individuals who work in these industries are often paid less than minimum wage and do not have access to health benefits or sick days. Many cannot afford to eat themselves. Jayaraman analyzes these challenges using a decade of research she has conducted. She argues that our food system will never be truly "sustainable" as long as it is sold and served under unsustainable working conditions.

The final part of the book, From Federal Policies to Local Programs: Solutions for a Sustainable Food System, offers suggestions for making our food system more sustainable. Susan Schneider calls for a "recasting" of agricultural law and policy. In her chapter, Schneider explains that an agricultural law (primarily the farm bill) that focuses on the economic vitality of agriculture as an industry has resulted in protectionist and exceptionalism laws and regulations. This approach has sidestepped broader discussions, including topics like the obesity crisis, soil conservation, and the inhumane treatment of animals. Schneider argues that agricultural law should be recast as the law of food, farming, and sustainability, with the sustainable production and delivery of healthy food to consumers as its central goal.

The next chapter offers an alternative to changing the farm bill as a way of achieving a sustainable food system. Jason J. Czarnezki focuses on the use of informal tools of regulation—such as eco-labels and informational regulation—to target individual behavior, as well as proposes structural changes to our food system. Some examples include creating better food system planning through state food policy councils and municipal planners, building on existing interests in intrastate and regional efforts supporting local food and local economies, and improving the management of existing alternative agricultural distribution and production systems.

Mary Jane Angelo offers solutions to a specific problem in industrial agriculture: pesticides. Mimicking the twelve-step program from Alcoholics Anonymous, Angelo offers twelve steps to breaking our pesticide addiction. For example, acknowledging an overreliance on the use of synthetically chemical pesticides is the first step (admitting you have a problem). Another example, looking to a higher power (step three), would include looking to how natural ecosystems function by integrating ecological resilience into sustainable agriculture.

Rather than focusing on a specific problem, in the final chapter, Jill Clark and Caitlin Marquis suggest a specific tool to address the larger question of achieving a sustainable food system: conducting a food policy audit. Using two case studies, one from Charlottesville, Virginia, and the other from Franklin County, Ohio, the authors explain the value and application of a food policy audit.

The complexity of our food system makes it difficult for a single book to offer a single solution for the creation of a sustainable food system. Instead this book offers perspectives from different stakeholders who offer their critiques of our food system and suggestions for moving forward. As a result, this book serves as a starting point for in-depth discussions on creating a sustainable food system in the twenty-first century.

I: The Elements of Our Complicated Food System
FOOD, LAND, AND FARMERS

1 Utopian Dream
A Farm Bill Linking Agriculture to Health
Marion Nestle, New York University

In the fall of 2011, I taught a graduate food studies course at New York University devoted to the farm bill, a massive and massively opaque piece of legislation then passed most recently in 2008 and up for renewal in 2012 (it was subsequently passed in 2014). The farm bill supports farmers, of course, but also specifies how the United States deals with such matters as conservation, forestry, energy policy, organic food production, international food aid, and domestic food assistance. My students came from programs in nutrition, food studies, public health, public policy, and law, all united in the belief that a smaller scale, more regionalized, and more sustainable food system would be healthier for people and the planet.

In the first class meeting, I asked students to suggest what an ideal farm bill should do. Their answers covered the territory: ensure enough food for the population at an affordable price; produce a surplus for international trade and aid; provide farmers with a sufficient income; protect farmers against the vagaries of weather and volatile markets; promote regional, seasonal, organic, and sustainable food production; conserve soil, land, and forest; protect water and air quality, natural resources, and wildlife; raise farm animals humanely; and provide farmworkers with a living wage and decent working conditions. Overall, they advocated aligning agricultural policy with nutrition, health, and environmental policy—a tall order by any standard, but especially so given current political and economic realities.

1. WHAT'S WRONG WITH THE CURRENT FARM BILL?

Plenty. Beyond providing an abundance of inexpensive food, the current farm bill addresses practically none of the other goals. It favors Big Agriculture over small; pesticides, fertilizers, and genetically modified crops over those raised organically and sustainably; and some regions of the country—notably the South and Midwest—over others. It supports commodity crops grown for animal feed but considers fruits and vegetables to be "specialty" crops deserving only token support. It provides incentives leading to crop overproduction, with enormous consequences for health.

The bill does not require farmers to engage in conservation or safety practices (farms are exempt from having to comply with environmental or employment standards). It encourages production of feed crops for ethanol. In part because Congress insisted that gasoline must contain ethanol, roughly 40 percent of U.S. feed corn is grown for that purpose, a well-documented cause of higher world food prices. Because the bill subsidizes production, it gets the United States in trouble with international trading partners, and hurts farmers in developing countries by undercutting their prices. Taken as a whole, the farm bill is profoundly undemocratic. It is so big and so complex that nobody in Congress or anywhere else can grasp its entirety, making it especially vulnerable to influence by lobbyists for special interests.

Although the farm bill started out in the Great Depression of the 1930s as a collection of emergency measures to protect the income of farmers—all small landholders by today's standards—recipients soon grew dependent on support programs and began to view them as entitlements. Perceived entitlements became incentives for making farms larger; increasingly dependent on pesticide, herbicide, and fertilizer "inputs"; and exploitative of natural and human resources. Big farms drove out small, while technological advances increased production. These trends were institutionalized by cozy relationships among large agricultural producers, farm-state members of congressional agricultural committees, and a Department of Agriculture (USDA) explicitly committed to promoting commodity production.

These players were not, however, sitting around conference tables to create agricultural policies to further national goals. Instead, they used the bill as a way to obtain earmarks— programs that would benefit specific interest groups. It is now a 357-page piece of legislation with a table of contents that alone takes up 10 pages. As the chief vehicle of agricultural policy in the United States, it reflects no overriding goals or philosophy. It is simply a collection of hundreds of largely disconnected programs dispensing public benefits to one group or another, each with its own dedicated constituency and lobbyists. The most controversial farm bill programs benefit only a few basic food commodities—corn, soybeans, wheat, rice, cotton, sugar, and dairy. But lesser-known provisions help much smaller industries such as asparagus, honey, or Hass avocados, although at tiny fractions of the size of commodity payments.

The 2014 bill organizes its programs into twelve "titles" dealing with its various purposes. I once tried to list every program included in each title, but soon gave up. The bill's size, scope, and level of detail are mind-numbing. It can only be understood one program at a time. Hence, lobbyists.

The elephant in the farm bill—its biggest program by far and accounting for about 80 percent of the funding—is SNAP, the Supplemental Nutrition Assistance Program (formerly known as food stamps). In 2015, as a result of the declining economy and high unemployment, SNAP benefits covered forty-six million Americans at a cost of $74 billion. In contrast, crop insurance costs "only" $9 billion, commodity programs $4 billion, and conservation about $6 billion. The amounts expended on the hundreds of other programs covered by the bill are trivial in comparison, millions, not billions—mere rounding errors.

What is SNAP doing in the farm bill? Politics makes strange bedfellows, and SNAP exemplifies logrolling politics in action. By the late 1970s, consolidation of farms had reduced the political power of agricultural states. To continue farm subsidies, representatives from agricultural states needed votes from legislators representing states with large, low-income urban populations. And those legislators needed votes from agricultural states to pass food assistance bills. They traded votes in an unholy alliance that pleased Big Agriculture as well as advocates for the poor. Neither group wants the system changed.

II. HEALTH IMPLICATIONS

The consequences of obesity—higher risks for heart disease, type 2 diabetes, certain cancers, and other chronic conditions—are the most important health problems facing Americans today. To maintain weight or to prevent excessive gain, federal dietary guidelines advise consumption of diets rich in vegetables and fruits. The 2014 farm bill has a horticulture title that includes organics, but aside from a farmers' market promotion program and some smaller marketing programs, does little to encourage vegetable and fruit production or to subsidize their costs to consumers. If anything, the farm bill encourages weight gain by subsidizing commodity crops that constitute the basic cheap caloric ingredients used in processed foods—soy oil and corn sweeteners, for example—and by allowing crop producers to use only 15 percent of their land to grow fruits and vegetables.

Neither human nature nor genetics have changed in the last thirty years, meaning that widespread obesity must be understood as collateral damage resulting from changes in agricultural, economic, and regulatory policy in the 1970s and early 1980s. These created today's "eat more" food environment, one in which it has become socially acceptable for food to be ubiquitous, eaten frequently, and in large portions.

For more than seventy years, from the early 1900s to the early 1980s, daily calorie availability remained relatively constant at about 3,500 per person. By the year 2000, however, available calories had increased to 4,200 per person per day, roughly twice the average need. People were not necessarily eating 700 more daily calories, as many were undoubtedly wasted. But the food containing those extra calories needed to be sold, thereby creating a marketing challenge for the food industry.

Why more calories became available after 1980 is a matter of some conjecture, but I believe the evidence points to three seemingly remote events that occurred at about that time: agriculture policies favoring overproduction, the onset of the shareholder value movement, and the deregulatory policies of the Reagan era.

In 1973 and 1977, Congress passed laws reversing long-standing farm policies aimed at protecting prices by limiting production. Subsidies increased in proportion to amounts grown, encouraging creation of larger and more productive farms. Indeed, production increased, and so did calories in the food supply and competition in the food industry. Companies were forced to find innovative ways to sell food products in an overabundant food economy.

Further increasing competition was the advent of the shareholder value movement to force corporations to produce more immediate and higher returns on investment. The start of the movement is often attributed to a 1981 speech given by Jack Welch, then head of General Electric, in which he insisted that corporations owed shareholders the benefits of faster growth and higher profit margins. The movement caught on quickly, and Wall Street soon began to press companies to report growth in profits every quarter. Food companies, already selling products in an overabundant marketplace, now also had to grow their profits—and constantly.

Companies got some help when Ronald Reagan was elected president in 1980 on a platform of corporate deregulation. Reagan-era deregulatory policies removed limits on television marketing of food products to children and on health claims on food packages. Companies now had much more flexibility in advertising their products.

Together, these factors led food companies to consolidate, become larger, seek new markets, and find creative ways to expand sales in existing markets. The collateral result was a changed society. Today, in contrast to the early 1980s, it is socially acceptable to eat in places never before meant as restaurants, at any time of day, and in increasingly large amounts—all factors that encourage greater calorie intake. Food is now available in places never seen before: bookstores, libraries, and stores primarily selling drugs and cosmetics, gasoline, office supplies, furniture, and clothing.

As a result of the increased supply of food, prices dropped. It became relatively inexpensive to eat outside the home, especially at fast-food restaurants, and such places proliferated. Food prepared outside the home tends to be higher in calories, fast food especially so. It's not that people necessarily began to eat worse diets. They were just eating more food in general and, therefore, gaining weight. This happened with children, too. National food consumption surveys indicate that children get more of their daily calories from fast-food outlets than they do from schools, and that fast food is the largest contributor to the calories they consume outside the home.

To increase sales, companies promoted snacking. The low cost of basic food commodities allowed them to produce new snack products—twenty thousand or so a year, nearly half candies, gum, chips, and sodas. It became *normal* for children to regularly consume fast foods, snacks, and sodas. An astonishing 50 percent of the calories in the diets of children and adolescents now derive from such foods. In adults and children, the habitual consumption of sodas and snacks is associated with increases in calorie intake and body weight.

Food quantity is the critical issue in weight gain. Once foods became relatively inexpensive in comparison to the cost of rent or labor, companies could offer foods and beverages in larger sizes at favorable prices as a means to attract bargain-conscious customers. Larger portions have more calories. But they also encourage people to eat more and to underestimate the number of calories consumed. The well-documented increase in portion sizes since 1980 is by itself sufficient to explain rising levels of obesity.

Food prices are also a major factor in food choice. It is difficult to argue against low prices and I won't—except to note that the current industrialized food system aims at producing food as cheaply as possible, externalizing the real costs to the environment and to human health. Prices, too, are a matter of policy. In the United States, the indexed price of sodas and snack foods has declined since 1980, but that of fruits and vegetables has increased by as much as 40 percent. The farm bill subsidizes animal feed and the ingredients in sodas and snack foods; it does not subsidize fruits and vegetables. How changes in food prices brought on by growth of crops for biofuels will affect health is as yet unknown but unlikely to be beneficial.

The deregulation of marketing also contributes to current obesity levels. Food companies spend billions of dollars a year to encourage people to buy their products, but foods marketed as "healthy"—whether or not they are—particularly encourage greater consumption. Federal agencies attempting to regulate food marketing, especially to children, have been blocked at every turn by food industries dependent on highly profitable "junk" foods for sales. Although food companies argue that body weight is a matter of personal choice, the power of today's overabundant, ubiquitous, and aggressively marketed food environment to promote greater calorie intake is enough to overcome biological controls over eating behavior. Even educated and relatively wealthy consumers have trouble dealing with this "eat more" environment.

III. FIXING THE FARM BILL

What could agriculture policies do to improve health now and in the future? Also plenty. When I first started teaching nutrition in the mid-1970s, my classes already included readings on the need to reform agricultural policy. Since then, one administration after another had tried to eliminate the most egregious subsidies (like those to landowners who don't farm) but failed when confronted with early primaries in Iowa. Embarrassed legislators ended direct payments in 2014, but found other methods for making sure that most benefits accrue to Big Agriculture. Defenders of the farm bill argue that the present system works well to ensure productivity, global competitiveness, and food security. Tinkering with the bill, they claim, will make little difference and could do harm. I disagree. The farm bill needs more than tinkering. It needs a major overhaul. My vision for the farm bill would restructure it to go beyond feeding people at the lowest possible cost to achieve several utopian goals:

Support farmers: The American Enterprise Institute and other conservative groups argue that farming is a business like any other and deserves no special protections. My NYU class thought otherwise. Food is essential for life, and government's role must be to ensure adequate food for people at an affordable

2 Land for Food in the Twenty-First Century

John Ikerd, University of Missouri Columbia

E nsuring access to enough farmland to meet the basic food needs of all will be a defining challenge of the twenty-first century. The sustainability of human life on earth, or at least human civilization as we know it, depends on the sustainability of global food production. Lester Brown of the Earth Policy Institute (EPI) identifies food scarcity as "the weak link" of modern society.[1] He points to the growing global demand for food and fuel, eroding soils, declining aquifers, and global climate change as major challenges to the future of human civilization. On the other hand, a 2009 United Nations (UN) High-Level Expert Forum on global food security reached quite different conclusions. While acknowledging the challenges of food scarcity, they concluded: "Overall, however, it is fair to say that . . . on a global scale there are still sufficient land resources to feed the world population for the foreseeable future, provided that investments required to develop these resources are made."[2]

The differences in conclusions reflect differences in the initial assumptions of the analyses. The EPI analysis assumed that rising energy costs, continued degradation of soil and water resources, and global climate change would make future increases in productivity impossible. It concluded that current negative trends, such as soil erosion and water depletion, must be reversed to ensure enough land for food. The UN analysis assumed that past increases in food production per hectare of farmland can continue in the future, although perhaps more slowly—given sufficient investments in research and production technologies. "New technologies to grow more from less land, with fewer hands"[3] was its solution to ensuring enough land for food.

Which conclusion is correct? The question of global food security is incredibly complex with so many interrelated and indeterminate factors that definite conclusions or precise predictions are conceptually impossible. Virtually all nontrivial questions of ecological, social, and economic sustainability encounter this same difficulty of complexity. Careful analysis of data and projections of trends obviously can inform individual perspectives on such issues. However, trying to determine whose science is right and whose is wrong or searching for middle ground among conflicting conclusions may be counterproductive.

Different conclusions arising from different assumptions often reflect different paradigms, which in turn reflect different worldviews. The validity of worldviews, or beliefs about how the world works, cannot be proven or disproven through science. Any discussion of whether there will be enough land for food must include a discussion of paradigms and worldviews. An example of particular relevance to questions of global food security and sustainability is the paradigm of neoclassical economics, which reflects a specific worldview.

I. THE PARADIGM OF NEOCLASSICAL ECONOMICS

The implicit assumption of the neoclassical economic paradigm is that human imagination, ingenuity, and creativity are capable of finding a substitute for any resource we may deplete or degrade and a technological solution to any problem we humans might create. This assumption is implicit in virtually all contemporary analyses of global food security. The scientific worldview that supports the neoclassical economic paradigm considers humans to be the supreme beings, and the earth simply an endless source of natural resources capable of sustaining infinite economic growth. The only significant scientific challenge in having enough of anything is to discover how best to manipulate nature to meet the needs and desires of humans. The only significant economic challenge to having enough is to ensure adequate economic incentives to develop new technologies and to allocate scarce resources among alternative uses.

To neoclassical economists, the projected slowing of gains in agricultural productivity in the UN report is simply a reflection of a lack of adequate economic incentives to develop and adopt more productive farming methods. As food becomes increasingly scarce and thus more economically valuable, the economy will allocate more resources to food production, there will be economic incentives for new production technologies, and the increasing food needs of a global society will be met. To agricultural scientists engaged in the EPI analysis, food production ultimately will be limited by the scarcity of productive natural resources.

The UN report embraces the economic assumption of the potential for unlimited growth in productivity in concluding that there will be enough land for food in the twenty-first century. The EPI analysis questions the assumption of limitless agricultural productivity but does not challenge the ability of markets to allocate natural resources to ensure enough land for food. We simply need to moderate growth in the global demand for food and mitigate the negative ecological challenges to growth in the global food supply. Eradicating global poverty was but one of four priorities included in the study for moderating future demand for food. Neither analysis seriously challenges the contemporary scientific worldview or the neoclassical economic paradigm.

Great civilizations of the past had developed highly sophisticated technologies, including complex irrigation systems and soil amendments, but the productivity of their agricultural lands eventually declined and their civilizations failed.[4] Their advances in production technologies failed to keep pace with the degradation of their natural resources and growing demand for food. Today's technologies are far more advanced than those of the past, but so is the degradation of the resources of nature and the human demands being placed on those resources.

The dominant thinking of today is essentially the same as in past failed civilizations. The earth exists for the sole benefit of humans. With adequate individual incentives we can extract and exploit its endless bounty. The focus of food production today, as in the past, is on natural resource efficiency and substitution with little consideration of natural resource degradation or growing economic and social inequities that threaten the sustainability of global food production. The future of humanity is being trusted to the same flawed thinking that failed great civilizations of the past.

II. THE CHALLENGE OF GLOBAL SUSTAINABILITY

When past civilizations collapsed, there were always civilizations elsewhere to carry humanity to a higher level of development or well-being. If today's global food system collapses, it would mean the end of humanity, or at least the end of human civilization as we have known it. In addition, global population has never before been remotely comparable to today's population, and global resources have never before suffered such severe ecological degradation. Soil, air, water, energy, climate, are all at risk of ecological collapse. Global society also is edging toward potential chaos, as the gap of economic and social disparity between the rich and poor grows ever wider both within and among nations. Six years after the global financial meltdown of 2008, the global economy remains on the verge of disintegration. Never before has the whole of humanity been on the verge of ecological, social, and economic collapse.

This unprecedented constellation of challenges has brought the essential question of *sustainability* to widespread public consciousness and has made it an essential watchword for all major corporations and government organizations. *How can we meet the needs of the present without diminishing opportunities for the future?* This is the fundamental question of sustainability. It seems intellectually reckless as well as ethically irresponsible to rely on the same ways of thinking that failed great civilizations of the past to meet the sustainability challenges of the future. The challenges of sustainability and global food security can be met, but not without a transformation in ways of thinking, including how we think about land for food.

Statistics regarding the prevalence of hunger vary widely—again, depending on assumptions. Relationships among food production, calorie consumption, and chronic malnutrition or hunger vary widely depending on the economic and social structures of the different countries of the world. According to a 2013 UN Food and Agriculture Organization (FAO) report of global food security, a total of 842 million people, or around one in eight people in the world, were estimated to be suffering from hunger, meaning they were not getting enough food to support active lifestyles.[5] The FAO report concluded that global hunger has fallen significantly in the past 30 years. However, global hunger statistics for years prior to 1990 vary widely and are widely questioned.[6] For example, estimates of global hunger during the late 1960s and early 1970s ranged from 500 million to 2 billion people, suggesting that hunger may have nearly doubled or may have been reduced by one-half over the past forty years.

The experts do seem to agree that high levels of global hunger persist, even though global agriculture currently produces enough food for everyone in the world. An FAO estimate indicates that global food production is sufficient to provide each person in the world with more than 2,700 calories per day.[7] This would be more than enough to ensure adequate food for everyone, if it were equally distributed among and within nations of the world.[8] With respect to sufficient land for food, an estimated 28 percent of the world's agricultural land area produces food that is lost or wasted.[9] The primary sources of food waste come from harvesting and storage losses in less developed countries and retail and consumer waste in developed countries. Most of this waste is avoidable.

Hunger today is not a consequence of a lack of land for food production but instead is a result of issues with distribution, access, and waste.

Hunger in the world today is avoidable or discretionary—not unavoidable or necessary. There is enough farmland in the world today to produce enough good food for everyone. However, many people of the world do not have access to enough good food produced on that land to meet their basic nutritional needs. People are hungry because the global economy depends too heavily on markets to provide food for the hungry, and many people are too poor to buy enough food to meet their needs. Admittedly, the agricultural production in some countries is simply not adequate to produce enough food for the entire population, even if food were equitably distributed. Such countries may be economically unable to import enough food to meet the basic food needs of their people. People in these countries are hungry because their entire nations are poor, and international food assistance programs are often inadequate to meet their food needs. Within such nations, poverty and hunger are unavoidable. Within the larger global society, however, poverty and hunger are avoidable or discretionary. Ensuring enough land for food in the future demands that we face the uncomfortable reality that hunger today is discretionary and unnecessary.

Most of the current discussions of land for food in the future focus on challenges of avoiding an absolute or unavoidable global scarcity of land for food, rather than the relative or discretionary scarcity of food from the land within and among nations. The former is a question of whether there is or will be sufficient land to produce enough food to meet the needs of all. The latter is a question of whether the land available not only will be adequate to meet the food needs of all but also whether the land will actually be used to meet the food needs of all. The former is a consequence of insufficient productivity of land resources; the latter is a consequence of lack of common access to the productivity of farmland. Distinctions between unavoidable and discretionary poverty and hunger are important in understanding the persistence of global hunger in the past, but even more important in addressing the twin challenges of poverty and hunger in the future.

III. THE REAL TRAGEDY OF THE COMMONS

The "tragedy of the commons" is an economic theory in which individuals acting independently and rationally, according to each one's economic self-interest, behave contrary to the long-term best interest of the group as a whole by depleting the productivity of some common resource.[10] The classic example focused on the logic of overgrazing common pasturelands by "economically rational" individual herdsmen. The purported means of avoiding the tragedy, by preventing overgrazing, was to enclose the commons, providing economic incentive for individual landowners to maintain the productivity of their individual plots. Markets could then allocate the use of land among individuals so as to sustain its long-run productivity, benefiting the group as a whole as well as the individual herdsmen.

Contrary to this economic theory, history has clearly demonstrated that enclosing land will not ensure its long-run productivity, and market economies will not allocate the use of farmland so as to provide food for all people in either present or future generations. Yet, few if any global food experts today seem willing to challenge the conventional economic thinking that markets will somehow allocate global land resources to provide food security for both present and future generations.

In his classic book, *The Great Transformation,* economist Karl Polanyi details the historical consequence of "commodifying" land and labor.[11] Prior to the "enclosure movement" of the sixteenth century, community groups had common access to land; it was not owned by individuals. Land was freely available for everyone to use to meet their basic needs for survival and sustenance, including their need for food. Uses of land for various purposes, including food production, were determined by community consensus, not by markets. *Unavoidable* or absolute poverty and hunger existed in many areas prior to the enclosures, as it does in some parts of the world today where land is held in common. However, *discretionary* or relational poverty and hunger only began when land was removed from the commons through enclosing it. The real "tragedy of the commons" was not that the agricultural commons were destroyed by overuse by self-seeking individuals but that removing land from the commons created a new kind of poverty and hunger.

Prior to the enclosures, market transactions were limited primarily to international trade. The primary means of meeting basic physical or material needs were through subsistence farming and local gifting economies. In gifting economies, goods and services are not bought or sold but instead freely given without explicit agreements concerning immediate or future rewards or reciprocity. Barter, or formal exchange of goods or services, was limited primarily to trade among people who did not know each other, and in some cases, people who were otherwise enemies. However, international markets had proven very effective in increasing the aggregate wealth of nations, as suggested by Adam Smith and other early capitalists, by allowing individual countries to capitalize on their economic comparative advantages. Enclosing the land, or the commons, would allow individuals to capitalize on their economic comparative advantages within and among villages and nations.

First, land had to be privatized and commodified so it could be bought and sold and thus reallocated among buyers and sellers in relation to their comparative advantages. The use of various parcels of land by specific individuals could then be determined by market competition, rather than community consensus. The commodification of land essentially forced the commodification of labor, as those without access to land for food were forced to sell their labor to employers, not only to thrive economically but to survive physically.

The fundamental problem of commodifying land and labor arises from the fact that people are inherently unequal in their ability to earn money in a market economy and thus have unequal amounts to spend in a market economy. We all have different physical and mental capacities, different aptitudes and opportunities, and different initial endowments of financial or physical resources. Nevertheless, all people in general have the same basic needs for food, clothing, shelter, and other necessities of life. The inequities of poverty and hunger are inevitable symptoms of market economies, which by their very nature reward people in relation to their contribution to the economy, not according to their needs. We cannot expect increasing global food production or reduced global food demand to alleviate hunger unless we address the global challenges of economic and social inequities.

Throughout history, significant numbers of people have been unable to meet even their most basic needs. Prior to the enclosures, the needs of such people were met by others in their communities who shared common access to the land. Following the initiation of enclosures in England in the 1500s, the government was forced to initiate a poverty program known as the English Poor Laws. These laws addressed the needs of indigent people who were left physically or mentally incapable of meeting their basic needs when communities lost common access to land. Although subsidized by taxes, these early laws were administered locally through village churches or charities.

In spite of growing poverty, enclosures of the commons continued during the seventeenth century: "the years between 1760 and 1820 [were] the years of wholesale enclosure in which, in village after village, common rights [were] lost."[12] During this period, the numbers of people impoverished or indigent exploded. In response, the English Poor Laws were nationalized and expanded in 1834 to cover the entire working class, not just the young, old, and physically or mentally disabled. For the first time, able-bodied workers were poor and hungry. This expansion of the Poor Laws was a clear indication that removing land from the commons had left even able, willing workers without means of providing enough food for their families.

Prior to the enclosures, the right to enough land to grow one's own food was long considered to be a fundamental right under what was known as "natural law." In 1690, John Locke proclaimed that land could be ethically removed from the commons only if "there is enough, and as good, left in the commons for others."[13] In comparing privatization of land to taking a drink from a flowing stream, he wrote, "[a]nd in the case of land and water, there is enough of both." Obviously, the good land left in the commons was often not enough to meet the needs of those who needed it most. To make matters worse, with land use or land tenure determined by its market value, land ownership was inevitably accumulated and consolidated among the wealthy, leaving even less land to meet the needs of the poor. In 1795, the American revolutionary Thomas Paine concluded: "[t]he landed monopoly ... has produced the greatest evil. It has dispossessed more than half the inhabitants of every nation of their natural inheritance ... and has thereby created a species of poverty and wretchedness that did not exist before."[14]

IV. THE POVERTY-HUNGER DILEMMA

In addition to the English Poor Laws, a variety of other attempts were made to protect the working class from the social upheaval triggered by removal of land from the commons. Nothing seemed to work. Thomas Paine proposed a universal, lifelong indemnity paid through the government to compensate the landless for their loss of access to the commons. He was reaffirming that if land belongs to anyone, it belongs to the people in common, and even if managed privately, it must still be used for the common good of all.

Paine's proposal was never implemented, but a variety of social welfare and food assistance programs were tried in the United States during the 1800s—with little success. Welfare programs were largely abandoned during the Gilded Age of the early 1900s. This left soup lines as the only alternative to starvation for many of the unemployed during the Great Depression of the 1930s. The New Deal social programs, initiated in the United States during the Great Depression years and expanded through the Great Society programs of the 1960s, were responses to the failure of the market economy to provide for the food, clothing, and shelter needs of the poor and hungry. Fascist governments turned to corporatist economies to meet their needs, with disastrous results for their people and the world. Other countries turned to socialism and communism in attempts to ensure that all people were capable of meeting their basic needs.

More recent food assistance programs in the United States have included free distribution of surplus agricultural commodities and the Food Stamp program, which evolved into today's Supplemental Nutrition Assistance Program or SNAP. While social welfare programs in the United States and elsewhere have undoubtedly mitigated the problems of hunger and poverty, none has adequately addressed the fundamental problem of economic inequities within market economies. Admittedly, the privatization of land and labor dramatically increased agricultural productivity and economic growth and brought many material benefits to humanity. However, market economies inevitably lead to economic inequities and have proven utterly incapable of eliminating poverty and hunger.

Communism and socialism have been frustrated by the same challenges as all social welfare programs dating back to the English Poor Laws. Many people only seem inclined to work when they have an individual incentive to do so. A common joke in the former Soviet Union was that "workers pretended to work and the government pretended to pay them." In the absence of economic incentives to be productive, people of such countries have suffered from unavoidable or absolute hunger. Whenever governments with market economies ensure people that their basic economic needs will be met regardless of whether they work, many simply choose not to work. Whenever governments make up the difference between what workers are able to earn in the labor market and earnings necessary to meet their basic economic needs, employers feel free to pay workers less, meaning governments must subsidize the workers more. As such practices persist, the rest of society eventually loses its willingness to pay the taxes needed to support such programs.

V. RECLAIMING THE COMMONS

Without access to land for food, significant portions of the populations, not just the young, old, or feeble, have been persistently unable to meet their basic needs for food. That said, simply returning land to the commons would not ensure enough land for food for all in the future. There are far too many people in the world to return to hunting and gathering or subsistence farming. However, restoring the basic rights of all people to benefit from the common wealth of nations remains a prerequisite for solving the twin problems of hunger and poverty. Ensuring enough good land to provide good food for all is but one dimension of ensuring equitable access to the common wealth of nations.

Industrial agriculture is not the answer to either hunger or poverty. It has produced more than enough food to meet the needs of all, but has utterly failed to address the fundamental problems of hunger and poverty. The allocation of land and food has been left to market incentives. As a result, people in industrial nations are consuming more calories than is consistent with good physical health, while those in developing nations lack sufficient calories to meet their basic physical needs. Within nations, those with higher incomes waste large amounts of food while those with lower incomes live on the verge of malnutrition. Even in the United States more people are hungry or "food insecure" today than during the 1960s, prior to the final phase in agricultural industrialization. More than 20 percent of children in the United States, one of the wealthiest countries in the world, face a persistent threat of hunger in food-insecure homes.[15]

Contrary to popular opinion, there is no clear evidence that hunger has been significantly alleviated in the so-called developing nations by the "Green Revolution." According to Vandana Shiva, a globally prominent ecologist and Indian food activist:

> The Green Revolution has been a failure. It has led to reduced genetic diversity, increased vulnerability to pests, soil erosion, water shortages, reduced soil fertility, micronutrient deficiencies, soil contamination, reduced availability of nutritious food crops for the local population, the displacement of vast numbers of small farmers from their land, rural impoverishment and increased tensions and conflicts. The beneficiaries have been the agrochemical industry, large petrochemical companies, manufacturers of agricultural machinery, dam builders and large landowners.[16]

The Green Revolution has had a similar effect as removing land from the commons, wherever its industrial farming practices have been employed. Lower commodity prices caused by increased production have depressed the market values, leaving subsistence farmers unable to compete for farmland or even buy the necessities they could not produce on their farms. Unable to continue farming, many families who were once reasonably well-fed remain unemployed, unemployable, and hungry in urban slums. The increased production on Green Revolution farms often finds its way into more profitable export markets, leaving the poor and hungry at home still poor and hungry. Forcing genetically engineered crops on farmers in developing countries and "land grabbing" by global corporations and foreign governments will only make the poverty and hunger situations worse in both developed and developing countries.

The ecological consequences of industrial agriculture have been well documented in the United States in terms of eroded and degraded soils, polluted streams and groundwater, depleted streams and aquifers, and the growing threat of global climate change. The socioeconomic consequences are apparent in the demise of independent family farms and the social and economic decay of rural communities, as the farms grow larger in size and fewer in numbers. In addition, basic human rights of self-determination and self-defense are systematically denied to rural residents who are forced to live with the clear and compelling public health risks associated with "factory farms."[17]

The fundamental problem is that industrial agriculture is driven by economic value—the pursuit of maximum profits. Economic value is inherently individual, instrumental, and impersonal. Since economic value is individual and instrumental, it is not economically rational to do anything for the sole benefit of someone else or for society in general. Since individual lives are inherently finite and uncertain, the economy places a premium on the present relative to the future. That is why people are willing to pay interest when they borrow money and expect to earn interest when they loan or invest money. It makes no sense to make investments if the returns will accrue to someone else in some future generation after the investor is dead. Industrial agriculture will not meet the needs of *all* in the present or leave equal or better opportunities for those of the future. It is not sustainable. Meeting the basic food needs of all, of both current and future generations, will require very different ways of thinking about agriculture and food.

VI. PUBLIC POLICIES TO ADDRESS UNAVOIDABLE OR ABSOLUTE HUNGER

Market economies are not able to and will not provide enough good food for all, and all previous attempts by governments at ameliorating this inherent deficiency of markets have failed to provide a sustainable solution. With respect to meeting the challenge of unavoidable or absolute scarcity of good farmland, specific parcels of naturally productive farmland could be identified and zoned for use to provide global food security. Enough land would need to be "permanently zoned" for food production to meet the basic food needs of current and future generations. Land zoned for food would also need to be sufficient in both quantity and quality to allow *sustainable* farming, in order to avoid further degradation and "mining" of productive farmland by *industrial* agriculture.

This approach would be very different from the current practice of buying "development rights" and placing land in agricultural "land trusts." Any speculative "development value" of land currently zoned for agriculture would be lost, rather than sold, but society has no responsibility to ensure the success of *specula-*

tive ownership of farmland. Owners of good farmland currently zoned for higher-valued economic uses could be compensated for "down-zoning" to agriculture, funded by taxing away speculative gains in lands that are up-zoned to higher-valued uses. Profits from up-zoning are essentially grants from society, as the owners of such land have done nothing productive to increase its value. Taxing away such profits would also remove any economic pressure to up-zone good farmland from agriculture to other uses. The same principle should be applied to all changes in land use. If the earth belongs to anyone, it belongs to everyone, equally and in common. In essence, farmland would be returned to the commons.

This could be done without depriving farmers of their right to benefit from their improvement of the land. Farming of land zoned for food could be treated as a public utility, as proposed by Willard Cochrane, secretary of agriculture during the Kennedy administration.[18] Existing owners of farmland would have limited "use-rights" to their land, which is all that private ownership of land was ever meant to entail. Would-be farmers could buy use-rights to farmland instead of buying land. Prices of farmland would reflect the economic value of land in producing food, not prices inflated by land speculation. Contracts for domestic food security purposes would be limited to family-sized farms, excluding farm corporations. As contractors of a public utility, farmers would be required to farm their land sustainably, and in return, would be ensured an income, above expenses, adequate to meet their basic economic needs.

In spite of claims to the contrary, sustainable farmers would be able to meet future global needs for food without significantly expanding the amount of land uses for food. Industrial agriculture advocates conveniently ignore that small farms already account for at least 70 percent of global food production, according to the United Nations Environmental Programme.[19] Other UN studies indicate that the production of such small farms could be more than doubled using sustainable farming methods rather than industrial agriculture. For example, a 2008 United Nations study of farming methods in twenty-four African countries found that organic or near-organic farming resulted in yield increases of more than 100 percent.[20] Another United Nations–supported study entitled *Agriculture at a Crossroads*, compiled by 400 international experts, concluded that agricultural production systems must change radically to meet future demand. It called for governments to pay more attention to small-scale farmers and sustainable farming practices.[21] With increased productivity, sustainable farmers could quite likely meet global food needs of the future without further expansion of land devoted to farming.

Sustainable farmers who produced enough to contribute more than their share to national food security could market their surpluses, giving them an economic incentive to increase the sustainable productivity of their land. Farmland not needed for domestic food security would also need to be farmed sustainably, but could be farmed by larger commercial operations to produce food for export to food-deficient nations. Markets would again be restricted to trade in agricultural surpluses, not depended on to provide domestic food security. All net farm income would be taxed, but not so heavily taxed as to remove all economic incentives for farmers to sustain the productivity of their land.

VII. PUBLIC POLICIES TO ADDRESS DISCRETIONARY OR RELATIVE HUNGER

Restoring land to the common wealth would not solve the problem of discretionary or relative hunger, which results from poverty rather than absolute food scarcity. However, the same basic principles could be followed in meeting the challenge of *relative* poverty as in meeting the challenge of *absolute* hunger. Solving the relative *poverty* problem would eliminate the discretionary *hunger* problem. Everyone in a given society could be provided with assurance of an income adequate to meet his or her essential economic needs, including enough good food to support healthy, active lifestyles. In return, everyone would be expected to contribute whatever he or she was able to contribute to the greater good of their society. To paraphrase Karl Marx, to each according to his or her need, from each according to his or her ability. Although contributions need not be economic, everyone would be expected to contribute to the extent of his or her ability.

This is not a radical new idea, although it may seem radical during these times of unprecedented economic and political inequity. Advocates of ensured "guaranteed minimum incomes" (GMIs) for all date back at least to Thomas Paine and Napoleon Bonaparte. They have spanned the political spectrum from the liberal U.S. senator Daniel Patrick Moynihan to conservative economist Milton Friedman.[22] The primary differences in opinion between liberals and conservatives seem to be whether a GMI should replace current supplemental income programs, such as Social Security and unemployment insurance, or whether a GMI should replace all income supplementing programs. With a bit of public pressure, such differences should be resolvable.

The commons in this case would be the common wealth of society, which was created and accumulated by those of past generations and now provides the cultural, social, and economic foundation for today's economy. Those who have benefited economically from having access to this common wealth have an ethical responsibility to invest a comparable portion of their economic dividends to sustain the common good of current and future generations. Their contribution must be adequate to ensure incomes for all sufficient to meet the basic economic necessities of life, including enough healthy and nutritious food to meet their physical needs.

Everyone would be provided with a basic education and the means of pursuing their chosen profession or means of contributing to the greater good of society, including higher education if necessary. In return, they would be expected to contribute their fair share to the greater good of their society, much as sustainable farmers would be expected to contribute their fair share to food security. Some might choose occupations such as painting, poetry, entertaining, and writing, which are important to the overall well-being of society but often fail to yield incomes sufficient to meet one's basic needs. Those who choose more economically lucrative occupations would still have economic incentives to do so. Whatever income anyone earned would need to be taxed in order to ensure that the economic needs of all are met, but not taxed so heavily as to remove all economic incentives to sustain economic productivity. Significant inheritance taxes, reaching down to middle-class wealth levels, would need to be restored to fund guaranteed minimum incomes as well as mitigate the intergenerational concentration of wealth.

The most difficult decision in implementing such programs to eliminate hunger and poverty would be in determining the fair shares that farmers and others must contribute to society in return for assurance that their basic needs will be met by society. The value of individual contributions cannot be determined by markets, as we have seen in the past failures of market economies to alleviate poverty and hunger. The contributions should instead be in relation to one's ability to contribute to the greater good of society, including social, ethical, and cultural contributions that have little or no economic value. Such decisions might be best left to local communities, where people have more personal knowledge of each other and can judge more fairly what others are able to contribute. Within communities, noneconomic contributions are more likely to be appreciated and responsibilities are less likely to be avoided.

Those who wanted to produce their own food should be ensured access to enough good land to meet the basic food needs of their family. The specific acreage of land provided would vary from community to community, depending on prevailing types of land, growing region, and other local factors. Those without land who wanted to farm for a living would be given an opportunity to "homestead" enough good land to make a living for their family, the acreage again varying from community to community. The family homesteaders would have to prove they could make a living from the land over some period of time, such as ten years, to claim title to use of the land. A portion of intergenerational transfers of farmland could be claimed by the community, as a land inheritance tax, to provide farmland for new farmers. Such a program would not only provide land for food but would also mitigate the concentration of farmland into large land holdings.

Returning the most important responsibilities for poverty and hunger to local levels would be much like returning to the days when land was held in common and people contributed whatever they could to ensure that the needs of all were met. Obviously, the current legal interpretations of land ownership would have to be reviewed and revised, returning to many of the historical principles of common property. Governments at state and national levels would need to be radically reformed, not only to ensure enough land for food but also to create and nurture a society and economy capable of meeting the economic needs of all of present generations without compromising opportunities for those of the future. However, the principle of subsidiarity, making decisions at the most local level possible, is a sound principle for economic and social responsibility. The foundation for a sustainable global economy always has been and always will be ensuring access to enough good land to provide good food for everyone, which means reclaiming the commons as part of the common wealth of nations.

NOTES

1. Lester Brown, Full Planet, Empty Plates: The New Geopolitics of Food Scarcity (2012).

2. Food and Agriculture Organization of the United Nations (FAO), *How to Feed the World in 2050*, http://www.fao.org/fileadmin/templates/wsfs/docs/Issues_papers/HLEF2050 _Global_Agriculture.pdf.

3. FAO *supra* note 2.

4. Jared Diamond, Collapse: How Societies Choose to Fail or Succeed (2005).

5. Food and Agriculture Organization of the United Nations (FAO), *Food Insecurity in the World: The Multiple Dimensions of Food Security* (2013), http://www.fao.org/publications /sofi/en/.

6. Ann Crittenden, *Food and Hunger Statistics Questioned*, N.Y. Times, Oct. 5, 1981, *available at* http://www.nytimes.com/1981/10/05/business/food-and-hunger-statistics-questioned .html.

7. Food and Agriculture Organization of the United Nations (FAO), *Agriculture and Food Security*, http://www.fao.org/docrep/x0262e/x0262e05.htm.

8. U.S. Department of Agriculture Report, *Estimated Calorie Needs per Day by Age, Gender, and Physical Activity Level*, http://www.cnpp.usda.gov/sites/default/files/usda_food_patterns/EstimatedCalorieNeedsPerDayTable.pdf.

9. Food and Agriculture Organization of the United Nations (FAO), *Food Wastage Footprint: Impacts on Natural Resources* (2013), http://www.fao.org/docrep/018/i3347e/i3347e.pdf.

10. Garrett Hardin, *The Tragedy of the Commons*, 162 Sci. (1968): 1243–48, *available at* http://www.dieoff.org/page95.htm.

11. Karl Polanyi, The Great Transformation: The Political and Economic Origins of Our Time (1944, 1957).

12. Edward Thompson, The Making of the English Working Class (1991), 217.

13. *Lockean Proviso*, Wikipedia, http://en.wikipedia.org/wiki/Lockean_proviso.

14. Thomas Paine, Agrarian Justice (1795), *available at* http://www.constitution.org/tp/agjustice.htm.

15. Alisha Coleman-Jensen, Mark Nord, Margaret Andrews, and Steven Carlson, USDA, Economic Research Report No. 125, *Household Food Security in the United States in 2010* (Sept. 2011), http://www.ers.usda.gov/publications/err-economic-research-report/err125.aspx.

16. Vandana Shiva, *The Green Revolution in the Punjab*, Ecologist, 21:2 (March–April 1991), *available at* http://livingheritage.org/green-revolution.htm.

17. Johns Hopkins Bloomberg School of Public Health, *Agriculture and Public Health Gateway*, Industrial Food Animal Production, http://aphg.jhsph.edu/?event=browse.subject&subjectID=43.

18. Richard Levins, Willard Cochrane and the American Family Farm (2000).

19. United Nations Environment Programme, Towards a Green Economy: Pathways to Sustainable Development and Poverty Eradication (2010), www.unep.org/greeneconomy.

20. United Nations Environment Programme, UNEP-UNCTAD Capacity-Building Task Force on Trade, Environment and Development, *Organic Agriculture and Food Security in Africa* (2008), http://unctad.org/en/docs/ditcted200715_en.pdf.

21. International Assessment of Agricultural Knowledge, Science and Technology for Development, *Agriculture at a Crossroads* (2009), http://apps.unep.org/publications/pmtdocuments/-Agriculture%20at%20a%20crossroads%20-%20Synthesis%20report-2009Agriculture_at_Crossroads_Synthesis_Report.pdf.

22. *Guaranteed Minimum Income*, Wikipedia, http://en.wikipedia.org/wiki/Guaranteed_minimum_income.

3

The Social Sustainability of Family Farms in Local Food Systems

Issues and Policy Questions

Jill K. Clark, Ohio State University; Shoshanah Inwood, University of Vermont; and Jeff S. Sharp, Ohio State University

I. INTRODUCTION

Much of the focus of creating sustainable local food systems has been on establishing alternative markets and on rebuilding regional food system infrastructure; while this structural approach is important, it overlooks the noneconomic motivations and decisions individual farmers and farm families make, which influences what is produced on the farm and whether to continue farming. Historically, the persistence of family farms over time has relied on intergenerational succession.[1] Succession, or transferring the management of the farm, is especially critical in those areas just outside of cities, also known as exurbia, as land markets are more active and competitive, and without a successful intergenerational succession of the farm, the potential of the land being sold to nonfarm interests increases. However, exurban farms contribute to U.S. agriculture, for example, by producing the majority of the nation's fruits and vegetables,[2] and can provide a potentially meaningful contribution to localizing the food system.[3]

In this chapter, we summarize years of our research with exurban farm families across the country, arguing that it is necessary to consider intergenerational succession to assess the long-term viability and sustainability of farm businesses engaged in sustainable food systems. We focus on exurban areas because they are the sites with the most varied and adaptive farms, and host the most organic and direct sales farms.[4] We also argue that our traditional understanding of intergenerational farm succession must be reconsidered due to the influx of new-comers to farming inspired by the opportunities of sustainable food systems, particularly in the area of local and direct sales. We refer to this new class of farmers as first-generation farmers (FGFs). While FGFs may be an exciting addition to some exurban landscapes, the permanence of these enterprises over the long term is a question that needs to be considered, particularly in comparison to the persistence of existing, multigeneration farm families (MGFs).

We focus on four topics related to the persistence of family farming in exurban areas and local food systems, with "local" or place-based characteristics of food systems merely being one aspect of sustainability. First, we briefly review the diversity of exurban farms and farmers (top box in Figure 1). We are particularly attentive to the diversity of motivations, personal values, and socialization of MGFs and FGFs involved in local food systems. Second, we speculate about how differences between MGFs and FGFs might lead them to manage intergenerational succession differently and how this process might impact the future farm enterprise viability in exurban areas (middle box in Figure 1). Third, we consider how the aggregated success or failure of the farm succession process can impact the exurban landscape as a whole. Our thinking related to the broader questions of landscape change is informed by insights from human ecology and the processes of invasion-succession[5] and land-use change (bottom box in Figure 1). We anticipate that because of differences between MGFs and FGFs, an increase of new FGFs in exurban areas may negatively affect the long-term persistence of family farming in these regions, particularly as these new farmers retire or bring their enterprise to a close.

Fourth, we close with a discussion of potential policy levers to use to ensure the sustainability of the exurban MGFs and FGFs. As U.S. policymakers and community groups turn their focus to food systems and the next generation of farmers, policy efforts should be considered to create the environment within which new farmers are cultivated, grown, and reproduced over time.

II. FARMING AND LOCAL FOOD SYSTEMS IN EXURBAN AREAS

The complexity and variability of exurban land markets and the indirect effects of urbanization result in more varied relationships between farms and

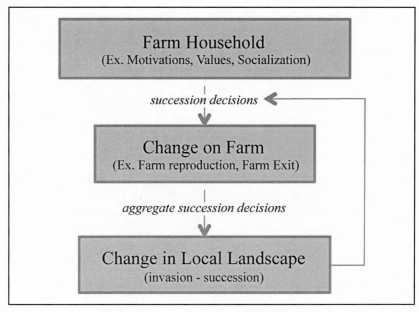

Figure 1. Dynamics of Farm and Landscape Succession

the agrifood system, and therefore produce a greater variety of farm types in these regions.[6] In part this is because exurban farms are uniquely positioned to participate in, and find market advantages in, local food systems.[7] Farmers adapt to urbanization in a few ways, resulting in (1) alternative enterprises, often small in size but with high value outputs engaging in nontraditional production; (2) recreational enterprises, often very small and operated by individuals best classified as hobby farmers; and (3) traditional enterprises, large operations engaged in conventional commodity production that because of land constraints, add an entrepreneurial business to grow the farm without expansion on the land base.[8]

In our work across the United States, we recognize two distinct types of farm families that are managing adaption in exurbia: multigeneration farm families and first-generation farm families. Previous research has found that some MGFs explore new livelihood and production strategies within a local food system in order to ensure the profitability and persistence of the farm operation. MGFs are attuned to the entrepreneurial opportunities in exurban areas, are looking for more value per acre, and are able to stack enterprises on a limited exurban land base.[9] The exurban setting also attracts new FGFs who aspire to be farmers and who take advantage of the urban-oriented market opportunities associated with serving local and alternative markets.[10] All this activity results in a mixed landscape with longtime farm families on farms ranging from a few acres (perhaps engaged in direct marketing fruits and vegetables or greenhouse production) to substantial ground (such as commodity grain production), as well as new farmers often operating on smaller acreages and engaged in alternative production and marketing and value-added strategies. While these enterprises coexist in the same landscape, the research referenced above suggests differing motivations for first-generation farm families and multigeneration farm families to engage in local food systems. It is the motivations, personal values, and socialization of FGFs and MGFs that contribute to future decisions about farm growth and succession. In reference to Figure 1, the next section focuses on the connection between the diversity found within farm households (top box) and farm change resulting from succession (middle box).

III. FARM SUCCESSION

Over time, in industrialized Western countries, decisions made by family farms have resulted in the continued persistence of the farm as both a production and social unit across many agricultural landscapes, including exurban areas. This persistence occurs in spite of the potentially lower returns from farming due to the high cost of land, low commodity prices, and competing nonfarm economies. Despite low returns, small and mid-sized farms have been able to remain economically competitive with large-scale farms due to the biological nature of farming, the flexible labor supply family farms offer, and the consistent ranking across many classes of farmers that believe farming and ranching, as a way of life, is of greater importance than profit.[11] Therefore, motivations and personal values must be considered as a primary influence on farm management decisions and farm adaptation.[12] In addition to motivations and values, socializing farm family children about the importance of passing the farming legacy on to the next generation is a powerful commitment that affects succession decision making.[13]

Inherently, all farmers must balance economic and noneconomic goals.[14] Examining the motivations exurban MGFs and FGFs have for farming, Inwood et al.[15] found discrete differences between these two groups. Multigeneration farmers and farm families were more likely to emphasize business strategies structured to ensure growth and income generation. The desire to pass the farm on to the next generation was a strong motivation to farm as profitably as possible in order for another generation to be incorporated onto the farm. While some MGFs have been part of local and regional food networks for multiple generations, others have only recently transitioned into local food production and sales directly to consumers because they viewed it as the only viable way to create new opportunities for the next generation preparing to inherit the farm. Many MGFs consciously choose not to sell their land for development despite the large returns that could be gained from selling to nonfarm interests. Some of the multigeneration farm household members interviewed in Inwood's study emphasized the importance of wanting their operation to be "a real family farm" and their desire to ensure opportunities for all family members.

In contrast, first-generation farmers and farm families were more likely to emphasize personal, spiritual, and environmental goals as motivations for land use and for being a part of a localized food system. However, while FGFs were trying to fulfill more ethereal motivations, they simultaneously acknowledged the importance of having a sufficient income that would ensure a certain standard of living (e.g., health insurance, retirement benefits, and college funds). To accomplish their personal and spiritual goals, many FGFs designed their farms with the primary objective of marketing their products through direct sales. Recognizing the limited income that could be made from farming, many sought to diversify their income by adding entertainment or educational activities to the farm enterprise and many continued to work (or have a spouse who worked) off the farm. Overall, these FGFs were less driven by income generation from the farm or the need to continue a family farm legacy compared to the MGFs.

For farm families, the heavy reliance on secondary income may be a double-edged sword. On one hand the ability to source capital may be more flexible, and additional income can buffer the farm enterprise from shocks and fluctuations in the market.[16] On the other hand, farmers with a nonfarm job often do not scale up their operations because of time constraints and/or lack of desire to scale up.[17] For FGFs, deriving their livelihood elsewhere could mean they are more flexible and can access capital in other ways, but it may also mean that in the future, they may just not be able to scale up because the more lucrative nonfarm job takes so much time and energy they have little desire to become any larger or to pass the farm on if it is not securing a livelihood. This same predicament can also affect MGFs that rely heavily on outside income and was documented in a study in central Ohio.[18]

Socialization can also have profound implications for farm persistence and adaptation.[19] The socialization process begins in early childhood, and in the case of family farm systems, is integral for developing a long-term commitment to the farm enterprise among heirs and nonfarming siblings.[20] People are socialized in a process in which dominant modes of thought and experiences are internalized and develop into a subconscious belief system.[21] Socialization, therefore, can influence motivations, goals, and values. Many MGFs are socialized to address the need to replicate family tradition, carry on farm legacies, or reproduce the family farm. This provides the basis for the "generation of practice or behavior" that can entrench farm households on particular paths, making adaptation to local food systems difficult because of a "path dependency" that prevents farm households from considering alternative production strategies.[22] While these beliefs can change over time, these changes tend to be slow. FGFs face a different set of challenges; many new farmers are older, with one-third of beginning farmers being over fifty-five years old[23] and most likely with older children. FGFs' demographics raise significant questions about their intentions of ever growing or sustaining their operation from generation to generation and the ability to socialize the next generation into farming as a way of life.

The importance of alternative off-farm experiences is also a component to the socialization process. Children who work on other farms and in other settings can play an important role in developing strategies that revitalize the farm business.[24] Couples or individuals who are brought up outside the agricultural industry are more likely to pursue less traditional and more egalitarian paths of business development and management.[25] For example, compared to conventional producers, farm families directly marketing their products had an operator or spouse who had influential experiences outside their rural roots.[26] Younger farmers, especially those new to farming, may be more entrepreneurial and willing to tolerate the risk associated with innovation because they are not restricted by previous investments in traditional farming assets.[27] Further, FGFs do not necessarily feel the need to reproduce the family farm and family tradition, which can often predetermine the kind of production that occurs on the farm.[28] Part of the expansion, development, and diversification of a farm operation can result from taking advantage of a future heir's off-farm work experience, knowledge, and skills either working on other farms and/or in other settings. These experiences bring new skills, insights, and innovations back to the farm that can increase the chance of successful farm reproduction that can benefit both MGF and FGF enterprises.[29]

As suggested earlier, during times of farm development or economic pressure on the farm, some farm families may be poised to take advantage of new local or direct markets because of their position in the family farm life cycle. The life cycle of the farm family may motivate farm growth or adaption to engage in local food systems and, therefore, is also an important influence on farm management, enterprise growth, and enterprise persistence.[30] Because farms in exurban areas are not all at the same point of their life cycle, not all farm families are in a mode for growth or succession.

One factor of the life cycle that is critical to farm expansion, diversification, or exit is the presence of a farm successor. In anticipation of succession, the enterprise prepares to support another family by engaging in growth behaviors that are marked by innovation and expansion as operators adopt new management strategies to grow the farm business. Farmers without a successor are more likely to reduce the activity of their enterprise and their enterprise mix.[31]

Entrepreneurial tendencies are most apparent when the family farm needs to restructure, and the ability of farm survival relies on the willingness of the farm family to adapt the operation to address these needs.[32] Highly diversified farms (with multiple production lines) present more opportunities for potential heirs to work and exert responsibility compared to highly specialized farms (with minimal production lines).[33] Therefore, for small and mid-sized farms, adapting to engage in local food systems can be a part of their reproduction strategies, shifting from commodity to direct markets and relationships with individuals. This strategy is particularly useful in exurban areas where stacking complementary enterprises on top of each other on the same land base makes more sense than trying to expand via incorporating more land into the operation.[34] If first-generation farmers and farm families do not have another generation to bring on the farm, the additional pressure to expand and build new farm-based businesses is absent, compared to the pressures multigeneration farmers and farm families face.

Related to succession, beginning farmers that are part of a multigeneration farm family most often have access to land and other resources, such as equipment and credit, to which first-generation farmers may not have access. MGFs often "pass down" wealth through the farming business;[35] this is particularly critical in exurban areas when land resources are typically limited. And yet, these resources can also present obstacles if the farm is heavily invested (and indebted) in a specific kind of production, making it potentially difficult for new farmers to do anything but jump on the family's technological treadmill. From a practical standpoint, FGFs generally start with few resources. Mailfert[36] found that FGFs struggled to mobilize materials and information resources; this phenomenon can be especially exacerbated in exurban areas due to the higher cost of land and services. Additionally, new farmers not from farm backgrounds sometimes lack production knowledge and advanced farming skills and experience a significant learning curve. These new entrants face a considerable learning curve as the biological nature of farming inherently requires a long time to master production skills. Therefore, newer entrants take longer to scale up production than more experienced farmers.

Given differences in motivations, personal values, socialization, and life cycles between multigeneration and first-generation farmers and farm families, we are left with some pressing questions. Will these FGFs turn into MGFs or do FGFs' differences prevent this shift? If FGFs do not become MGFs, will a new crop of FGFs replace exiting FGFs to continuously perpetuate the FGF position on the landscape? Or will FGFs' farms only exist for their lifetime? At what point will FGFs outnumber MGFs if MGFs fail to persist on the landscape? Questions about individual farm persistence across generations lead us to the next section, which examines more fundamental questions about farming landscape persistence as a whole.

IV. LANDSCAPE PERSISTENCE: INVASION-SUCCESSION

We turn to the concepts of invasion and succession to anticipate and understand larger scale production patterns and the ways land can continue to remain in agricultural production even when it is farmed by a new family. The invasion-succession model helps explain the way one population supplants another one and one social system is replaced by its predecessor; the invasion-succession model has frequently been used by sociologists and geographers in the context of neighborhood change.[37] Invasion-succession explains how a new regime, or community, can come to occupy or succeed an established community. When a new group arrives in an area, it is possible the new group may eventually become dominant in that area. For example, if the new group grows sufficiently large on a particular landscape, a tipping point may be reached that causes the original group to leave. However, it is not always the case that the new group replaces the previous group. The new group can assimilate, and the two can coexist. Alternatively, the new group may be driven out by the established group. Although human ecology has been criticized for failing to account for the political economy,[38] the concepts of invasion-succession could be applied to the exurban agricultural landscape to ask new types of questions about the nature of farm persistence in these areas and the competition and prevalence of different types of farmers on this landscape.

If the local food system movement is attracting new farmers to engage in production to serve nearby urban markets, a question that might be considered is the extent to which farmers serving these local markets might come to dominate the landscape. Or will these new farmers exploit niches underutilized by established farmers and come to coexist with other exurban farmers and residents? Or will new farmers emerge on the landscape, but then fade over time as established farmers adapt to new market opportunities and absorb the newer farm enterprises as they fail to reproduce themselves? An important component to this question is the frequency with which these new types of farmers will be established, reproduce themselves across time, and possibly expand or proliferate to eventually dominate the landscape. In other words, will the FGF invaders assimilate and become MGFs, or will these new farmers sell to a new FGF family or a nonfarmer in the future, enabling a new class of invader? The implication is that FGFs' next-generation decisions can affect the balance of family farms and the future ability of nonfarmers to succeed these FGFs on the landscape. Alternatively, is there an opportunity for FGFs, MGFs, and nonfarmers to coexist, creating a landscape of mixed enterprise types in a tighter and more competitive exurban landscape? We attempt to address these questions later in this chapter.

To visualize how the exurban landscape might look as the result of varying outcomes of invasion by new farm enterprise types, Figure 2 presents four simplified landscapes for discussion purposes. Figure 2a is an illustration of an initial landscape occupied by large farms. For the purposes of this chapter, these "large" farms are considered to be conventional farms serving conventional national and international markets. They are large compared to the small farms serving alternative or local markets (as shown by the smaller squares). In the scenario shown in Figure 2b, due to urbanization pressures and the lack of farm heirs, some of the large farms are fragmented into smaller farms and residential lots. The remaining large farms may incorporate some elements of local food system–related activities into their production. Enough large farms persist to support some historic commodity infrastructure, and the smaller farms are adapting to the joint conditions of a fragmented landscape and the exurban market. In Figure 2c, all the large farms have been fragmented and now are smaller local food system–oriented farms and nonfarm residents (in gray). Between Figure 2b and 2c, a tipping point is reached where conditions like a tight land market, lack of agricultural services and infrastructure, and a lack of heirs could lead to all large farms being sold (in addition, parcel fragmentation and increasing scale of equipment makes large-scale agricultural production difficult). Figure 2c shows a landscape where the competition is for smaller lots between small farmers and nonfarmers, no longer between large and small farmers and nonfarmers. After a generation or so, one can imagine reaching another tipping point between Figure 2c and 2d, where a few farmers engaged in local food systems are either still on the land or have passed the farm on to an heir or sold it to another farmer, but where most sold their small farms to nonfarmers.

The tipping point between Figures 2b and 2c harkens back to debates a few decades ago about the impermanence syndrome. Berry[39] described the impermanence syndrome as the conditions created by an exurban environment that cause farmers to anticipate that their land will eventually be converted from agricultural to urban uses. Then, the increasing difficulties associated with farming around a growing nonfarm population lead to a gradual disinvestment and then exit from farming. The difference between the impermanence syndrome and what we suggest here is that the lack of farm succession (lack of heirs for MGFs and no plans for succession with FGFs) plays a greater role in the whole scale conversion of the exurban landscape to nonfarming uses than the impermanence syndrome. This potential conversion from lack of farm succession creates a serious bottleneck for local and regional food system development that will be difficult to overcome if it is not addressed early.

V. FINDING A BALANCE

The successful establishment of new farm enterprises, the successful growth of existing farms, and the successful reproduction of farms across generations all impact the long-term productivities and viability of exurban agriculture and the future of local food systems. Therefore, a key variable affecting both agriculture in exurban areas and local food system efforts is the question of who will be the next generation of farmers in light of fewer farm heirs returning to the farm and the relatively small number of beginning farmers. Recognizing that the "future of U.S. agriculture depends on the ability of new generations to establish successful farms and ranches," the USDA has launched a number of grant and program initiatives to support new and beginning farmers (individuals who have farmed less than ten years) by addressing access to farmland and capital, succession, tenure, and stewardship.[40] In addition, the USDA Know Your Farmer Know Your Food campaign was created to promote market development for small and medium farms by linking them more directly with consumers. Our research is a direct response to these efforts and questions the role of both new entrants with multigeneration farm backgrounds and those without farm backgrounds in influencing the structure of the exurban agrilandscape.

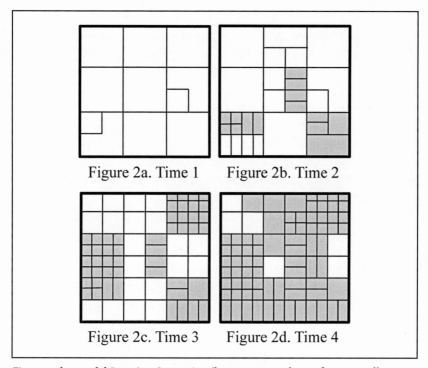

Figure 2a. Time 1   Figure 2b. Time 2

Figure 2c. Time 3   Figure 2d. Time 4

Figure 2a, b, c, and d. Invasion-Succession (large squares = larger farms; small squares = smaller farms; gray lots = residential lots)

In this chapter we speculated that the differences between multigeneration farm families and first-generation farm families that are drawn to the local food system movement have direct implications for farm enterprise reproduction and the long-term persistence of working agricultural landscapes in exurban areas. An increase of new farmers operating in exurban areas (especially those that fail to reproduce) may negatively affect the balance achieved by the persistence of the family farms in exurban areas in the past. This imbalance may result in the weakening of farm stability for the next generation unless we shift our thinking and outreach programming.

Perceiving problems associated with the conventional food system, first-generation farmers and farm families often consciously choose to engage in alternative farming practices, selling their products through local food markets to achieve personal needs and establishing entrepreneurial enterprises in the exurban land market. Because the need to replicate the family farm tradition may not be as integral to the operation, FGFs may have more flexibility in deciding how to initially develop their enterprise. In addition, many of them bring off-farm experiences that can contribute to entrepreneurial endeavors. But these same off-farm experiences together with the lack of farming socialization can lead to less on-farm knowledge and access to resources, capital, and otherwise. Further, many of these farmers may not be motivated to attempt an intrafamily farm transfer to maintain the business they have built. One possibility is that these farms will be in constant flux, with constant entrants and exits. But in an exurban landscape, land prices present a significant barrier. And there is a greater chance that the land will go to nonfarming heirs, with land retained for purely residential uses.

Some multigeneration farm families view strategies like direct sales as a way to grow a business on a small amount of ground or in highly competitive land markets. They emphasize vertical growth on the farm to create opportunities for the next generation, taking advantage of life cycle differences within the family to fill production, marketing, and childcare/household needs. MGFs are most often motivated to pass on the farm, but fewer and fewer children are returning to the home farm.

These considerations lead us to a more important question: What are the options when fewer farm heirs return to the farm or when first-generation farmers do not actively seek succession of their farm? Programs in the United States, such as FarmLink, have focused primarily on land transfers and less on the educational and startup needs of young and beginning farmers. Further, these programs are not designed to build the culture of succession with new farmers. Another model that may be more applicable, particularly in exurban locations, is one developed by the French government. State support for beginning farmers has a long history in France, dating back to the 1950s.[41] The French define beginning farmers by age, under forty years old (rather than by years farming as USDA does), and have developed a series of programs and incentives to recruit and retain young farmers through educational opportunities, preferential loans, and tax and social security benefits. In addition, recognizing that more beginning farmers are coming from nonfarm backgrounds, the program has evolved to account for the unique needs of this subset of beginning farmers. The program now offers incentives for farmers exiting agriculture, without an heir, to transfer their land to a candidate from a nonfarming background.[42]

In the United States, much attention has been paid to beginning farmers, especially since the passage of the 2008 Farm Bill.[43] As the United States prepares for one of the largest farm transitions in its history, it may be prudent for the USDA to reconsider the lens through which national farm policy is developed by taking a cue from the French model, which focuses not just on the startup of such businesses, but their persistence over time.

Resources, such as financing and land acquisition, and training, on topics such as production practices, marketing, financial planning, and business planning and management, are often covered in beginning farmer programs.[44] Hamilton[45] provides several suggestions on how to make these resources and trainings more accessible. One suggestion encourages the USDA to create "new farmer agents." These agents will be the personal portal to federal opportunities, such as those offered through the Farm Service Agency, Natural Resources Conservation Service, and, we add, university Cooperative Extension. Because these new farmer agents would be familiar with and embedded in the community, one could even conceive them being the portal to other community resources. Further, Hamilton suggests that these new farmer agents be networked throughout the United States with other agents to ensure effective means of federal priorities and initiatives reaching local communities.

The inclusion of long-time farmers in farmer learning communities could be supported through a "New Farm Nurturer" award.[46] The USDA could partner with other institutions and businesses (e.g., input and equipment dealers) that rely on new farmers to create an awards program akin to other farmer awards programs such as "outstanding farmer" or "top corn grower."[47] The award would recognize farmers who participate in land-link programs or enter into a long-term lease with a new farmer. This reward could be based on practice-based mentoring that could further socialize first-generation farmers to consider the long-term sustainability of their farm. Finally, the award could be offered through the "new farmer agent" network.[48]

At the local level, new farmer agents could engage more broadly with the community through such initiatives as civically oriented local food policy councils. As public interest in local food increases, food policy councils are being developed to address this interest though local food system initiatives by acting as conveners of stakeholders from across the local food system. These initiatives include setting policy agendas, focusing on low-income consumers, and building infrastructure between small and mid-size farmers and their consumers. New farmer agents could interface with these councils, working together to ensure the local policy environment (such as land use planning and zoning) is conducive to not just growing, but sustaining new farms over time. Food policy councils could be the broader connection between new farmers and other components of the local food system.[49]

Another consideration is to develop strategies to transition FGFs to be MGFs, essentially socializing FGFs to consider the life of their farm beyond their management, regardless of whether their successor is an heir. Establishing farms is one obstacle, but scaling up and sustaining these investments across time is even more complicated, as we have seen especially when new farmers differ in socialization, motivations, and long-range plans. Niewolny and Lillard[50] suggest that practitioners consider facilitating beginning farmer learning communities using strategies such as study circles, community forums, or collaborative leadership development that are culturally accessible to new farmers. Social networking is a necessary means by which farmers learn.[51] Participants in these learning groups may also be new local community members. New entrants to agriculture, who are intentionally choosing a specific way of farming, may exhibit attitudes and behaviors different from the local farming community they settle in. As such, practitioners should seek out multigeneration farmers that are particularly open to new methods of production and to work with first-generation farmers, with the objective of instilling a long-term community commitment and culture of succession.

For a long time, many exurban areas have experienced a balance in invasion-succession of farm enterprises. As the anchors, multigeneration farmers and farm families have enabled the farm landscape to persist. If a growing proportion of MGFs choose not to farm, the viability and sustained invasion by FGFs will tip the landscape structure in favor of FGFs. At first no discernable differences may be apparent on the landscape. However, a delayed effect may arise in the next generation as (1) FGFs may choose not to pass on the farm business or find themselves without a mechanism to do so, and (2) fewer MGFs are able to execute a successful succession to the next generation. Indeed, it is the "family" farm in exurban areas that may be in jeopardy in that the farm may be just a one-generation FGF operation. Given this possibility, the most desirable, balanced exurban localized food production landscape might be found in Figure 2b. Figure 2b represents an exurban landscape that has elements of individual farm stability and landscape stability struck through the balance of both large and small FGF and MGF farms serving local and global markets.

In this chapter, we have begun to explore issues related to the future of the exurban farm landscape through a local food systems lens. If first-generation farm families continue to be attracted toward local food systems, these landscapes could experience even more instability in the future than anticipated.

Therefore, local food system development and new farmer cultivation efforts must include not only infrastructure and startup costs but also support for the culture of farm persistence and replication.

NOTES

The authors would like to acknowledge the contributions to this research by Molly Bean Smith and Douglas Jackson-Smith. Research funding was provided by USDA National Research Initiative Grant # 2005-35401-15272.

1. John W. Bennett, ed., Of Time and the Enterprise: North American Family Farm Management in a Context of Resource Marginality (1982); Ruth Gasson and Andrew Errington, The Farm Family Business (1993); S. Salamon, Prairie Patrimony: Family, Farming, and Community in the Midwest (1992).

2. Shoshanah M. Inwood and Jill K. Clark, *Farm Adaptation at the Rural-Urban Interface*, 4 J. Agric., Food Sys., & Cmty. Dev. (2013).

3. Kate Clancy and Kathryn Ruhf, *Is Local Enough? Some Arguments for Regional Food Systems*, 25 Choices (2010).

4. Christopher R. Bryant and Thomas R.R. Johnston, Agriculture in the City's Countryside (1992); Inwood and Clark, *supra* note 2.

5. Kent Schwirian, *Models of Neighborhood Change*, 9 Ann. Rev. Sociol. (1983).

6. Bryant and Johnston, *supra* note 4.

7. *Id.*; Ralph Heimlich and Douglas Brooks, *Metropolitan Growth and Agriculture: Farming in the City's Shadow* (1989).

8. *Id.*; Shoshanah M. Inwood and Jeff S. Sharp, *Farm Persistence and Adaptation at the Rural Urban Interface: Succession and Farm Adjustment*, 28 Rural Stud. (2012).

9. G. A. Alsos et al., *Farm-based Entrepreneurs: What Triggers the Start-up of New Business Activities?*, 10 J. Small Bus. & Enter. Dev. (2003); Inwood and Sharp, *supra* note 8.

10. Carla Barbieri and Edward Mahoney, *Why Is Diversification an Attractive Farm Adjustment Strategy? Insights from Texas Farmers and Ranchers*, 25 J. Rural Stud. (2009); Jill K. Clark et al., *Community-level Influences on Agricultural Trajectories: Seven Cases across the Exurban U.S.*, *in* Geographical Perspectives on Sustainable Rural Change (Richard G. Winchell et al., eds., 2010).

11. Frederick H. Buttel et al., The New Sociology of Agriculture, I: The Political Economy and Social Structure of Farms, Farm Households and Farm Labor (1990); E. P. Durrenberger and S. Erem, Anthropology Unbound: A Field Guide to the 21st Century (2007); Nola Reinhardt and Peggy Barlett, *The Persistence of Family Farms in United States Agriculture*, 29 Sociol. Ruralis (1989).

12. G. Coleman and S. Elbert, *Farming Families: The Farm Needs Everyone*, 1 Research in Rural Sociol. & Dev. (1984); Ruth Gasson, *Goals and Values of Farmers*, 24 J. Agric. Econ. (1973); Gasson and Errington, *supra* note 1; Matt Lobley and Clive Potter, *Agricultural Change and Restructuring: Recent Evidence from a Survey of Agricultural Households in England*, 20 J. Rural Stud. (2004); Salamon, *supra* note 1; Mark Shucksmith and Vera Herrmann, *Future*

Changes in British Agriculture: Projecting Divergent Farm Household Behaviour, 53 J. Agric. Econ. (2002).

13. Bennett, *supra* note 1; Gasson and Errington, *supra* note 1; N.C. Keating and H.M. Little, *Choosing the Successor in New Zealand Family Farms,* 10 Fam. Bus. Rev. (1997); L. Kennedy, *Farm Succession in Modern Ireland: Elements of a Theory of Inheritance,* 44 Econ. Hist. Rev. (1991); Salamon, *supra* note 1; J.E. Taylore et al., *Succession Patterns of Farmer and Successor in Canadian Farm Families,* 63 Rural Sociol. (1998).

14. Gasson and Errington, *supra* note 1.

15. Shoshanah Inwood et al., *The Differing Values of Multigeneration and First-Generation Farmers: Their Influence on the Structure of Agriculture at the Rural-Urban Interface,* 78 Rural Sociol. (2013).

16. Reinhardt and Barlett, *supra* note 11; Shucksmith and Herrmann, *supra* note 12.

17. J.S. Sharp and M.B. Smith, *Farm Operator Adjustments and Neighboring at the Rural-Urban Interface,* 23 J. Sustain. Agric. (2004).

18. *Id.*

19. Bennett, *supra* note 1; Gasson and Errington, *supra* note 1; Keating and Little, *supra* note 13; Kennedy, *supra* note 13; Salamon, *supra* note 1; Taylore et al., *supra* note 13.

20. Donald J. Jonovic and Wayne D. Messick, Passing Down the Farm: The Other Farm Crisis (1986); Salamon, *supra* note 1; Joel Salatin, Family Friendly Farming: A Multi-Generational Home-Based Business Testament (2001).

21. Shucksmith and Herrmann, *supra* note 12.

22. *Id.*

23. Mary Ahearn and Doris Newton, USDA, *Beginning Farmers and Ranchers* (2009).

24. Jonovic and Messick, *supra* note 20.

25. Ruth Gasson et al., *The Farm As a Family Business: A Review,* 39 J. Agric. Econ. (1988).

26. Andrew W. Gilg and Martin Battershill, *The Role of Household Factors in Direct Selling of Farm Produce in France,* 90 Tijdschrift voor Economische en Sociale Geografie (1999).

27. Barbieri and Mahoney, *supra* note 10.

28. Clark et al., *supra* note 10.

29. Jonovic and Messick, *supra* note 20.

30. Bennett, *supra* note 1; Gasson and Errington, *supra* note 1; Clive Potter and Matt Lobley, *Aging and Succession on Family Farms: The Impact on Decision-Making and Land Use,* 32 Sociol. Ruralis (1992).

31. Bennett, *supra* note 1; Mieke Calus et al., *The Relationship between Farm Succession and Farm Assets on Belgian Farms,* 48 Sociol. Ruralis (2008); Gasson and Errington, *supra* note 1; Sarah Johnsen, *The Redefinition of Family Farming: Agricultural Restructuring and Farm Adjustment in Waihemo, New Zealand,* 20 J. Rural Stud. (2004); Potter and Lobley, *supra* note 30; A.M. Stiglbauer and C.R. Weiss, *Family and Non-family Succession in the Upper-Austrian Farm Sector,* Cahiers d'Économie et Sociologie Rurales (2000).

32. Johnsen, *supra* note 31.

33. Gasson and Errington, *supra* note 1.

34. Inwood and Sharp, *supra* note 8.

35. Gasson and Errington, *supra* note 1.

36. Kate Mailfert, *New Farmers and Networks: How Beginning Farmers Build Social Connections in France*, 98 Tijdschrift voor Economische en Sociale Geografie (2006).

37. Schwirian, *supra* note 5.

38. M. Gottdiener, The Social Production of Urban Space (1994).

39. David Berry, *Effects of Urbanization on Agricultural Activities*, Growth and Change (Jul. 1978).

40. USDA Cooperative State Research, Education and Extension Service, Family Farm Forum (Apr. 2008).

41. Mailfert, *supra* note 36.

42. *Id.*

43. N. D. Hamilton, *America's New Agrarians: Policy Opportunities and Legal Innovations to Support New Farmers*, xxiii Fordham Envtl. L.J. (2011); K. L. Niewolny and P. T. Lillard, *Expanding the Boundaries of Beginning Farmer Training and Program Development: A Review of Contemporary Initiatives to Cultivate a New Generation of American Farmers*, 1 J. Agric., Food Sys., & Cmty. Dev. (2010).

44. Niewolny and Lillard, *supra* note 43.

45. Hamilton, *supra* note 43.

46. *Id.*

47. *Id.*

48. *Id.*

49. *Id.*

50. Niewolny and Lillard, *supra* note 43.

51. *Id.*

4 Achieving Social Sustainability of Food Systems for Long-Term Food Security

Molly D. Anderson, Middlebury College

S ocial sustainability receives less attention in the United States than environmental and economic sustainability, and its meaning is still ambiguous. It is frequently equated with social justice or equity, but what does this really mean? What kind of justice or equity? For whom? Achieved in what ways? At what costs and with what trade-offs? Just as the term *social sustainability* remains poorly articulated in general, its meaning in food systems is not well defined.

Food security and meeting other basic needs are usually considered to be part of the social sustainability of food systems. A food system where people are food insecure is hardly sustainable. A common definition of food security is the one used by the U.S. Department of Agriculture: "access by all people at all times to enough food for an active, healthy life." How social sustainability is defined and operationalized makes a big difference in the answers to questions of who will enjoy food security, for what period of time, and under what terms. Food systems can provide many additional goods and services as well as food security (decent jobs at living wages, cultural amenities, watershed protection, nutrient cycling, etc.). Aspects related to wages, working conditions, and community relations are usually considered to be part of social sustainability, at a minimum, and these may help to determine food security.

This chapter describes how social sustainability is being addressed in the U.S. food system now through voluntary standards systems, principles, and state and municipal food plans. The United States has not defined social sustainability in food systems at the national level, and most efforts to operationalize social sustainability in food systems do not link it with economic, social, and cultural rights. While there are many reasons for this absence, public perceptions of agriculture play a part. This chapter traces connections between social sustainability concepts and agrarianism in the United States. Agrarian literature has linked democracy and wholesome family values with small-scale family farming, and has helped to feed contemporary enthusiasm for buying local food and supporting local farmers. Gaps in the agrarian lineage may have contributed to the current lack of emphasis on social sustainability and its connections with human rights in the United States. This chapter argues that a rights-based approach and greater attention to governance, in particular the need for democratic decision making that includes the voices of vulnerable and marginalized people in the food system, are essential aspects of social sustainability but are underemphasized in the United States and in our shared agrarian perspective, despite growing global consensus on their importance. Without these components of social sustainability, widespread food security will be very difficult to achieve.

I. SOCIAL SUSTAINABILITY IN INTERNATIONAL VENUES

Many sets of standards, indicators, and principles are emerging internationally that include social sustainability and have relevance to food systems, although often with much broader applicability. For example, in June 2008, the International Labor Organization (ILO) adopted a Declaration of Social Justice for a Fair Globalization, articulating its commitment to a Decent Work agenda, including criteria for employment, social protection, social dialogue, and rights at work. This is only the third revision of the ILO's principles and policies since it was founded in 1919. The most recent version drew from and extended the basic labor rights that had been agreed to in the 1998 ILO Declaration on Fundamental Principles and Rights at Work: freedom of association and the right to collective bargaining, the elimination of all forms of forced or compulsory labor, the abolition of child labor, and the elimination of discrimination in employment and occupation.[1]

Several other United Nations (UN) agencies have developed standards, indicators, or principles to inform the interface of businesses and workers. The Global Compact has established ten basic principles for businesses that include human rights, labor, and anticorruption, in addition to environmental responsibility. These are designed to ensure that markets, commerce, technology, and finance benefit economies and societies everywhere. In 2012 at the UN Rio+20 Conference, member states reiterated the importance of corporate sustainability reporting in the context of sustainable economic growth. They called on the UN to assist industry, interested governments, and relevant stakeholders to develop models for best practice. The UN secretary-general appointed the UN Conference on Trade and Development (UNCTAD), which deals with fairness in trade and investment and harmonizes guidelines on corporate social responsibility, as an official implementing body for this call to action. UNCTAD's work has clear applications in food systems, particularly given documentation of ways that trade and investment practices have disadvantaged small-scale growers.[2]

Standards, indicators, and principles developed by the Food and Agriculture Organization (FAO) deserve special mention since this is the UN agency charged with food security. The first of its three main goals is the eradication of hunger, food insecurity, and malnutrition.[3] FAO created a draft framework for sustainability in food systems in 2009, mapped and compared relevant indicators, reached out to stakeholders through e-consultations and other convenings, and piloted draft guidelines for Sustainability Assessment for Food and Agriculture (SAFA) in 2012–2013. The guidelines were finalized in 2013 and FAO is now developing guidelines specifically for small-scale farmers who grow on less than twenty-five acres.[4] SAFA includes indicators within the "social well-being dimension" on the themes of human health and safety, equity, cultural diversity, labor rights, fair trading practices, and decent livelihoods. The indicators for the themes are grouped into the right to quality of life; wage level; capacity development; fair access to means of production; fair pricing and transparent contracts; rights of suppliers; employment relations; forced labor; child labor; freedom of association and right to bargaining; nondiscrimination; gender equality; support to vulnerable people; safety and health trainings; safety of workplace, operations, and facilities; health coverage and access to medical care; public health; indigenous knowledge; and food sovereignty.[5] These indicators are explicitly linked to the 1948 Universal Declaration of Human Rights, the first international agreement on human rights. The other dimensions of the SAFA, in addition to social well-being, are environmental integrity, economic resilience, and good governance. Identification, engagement, and effective participation of stakeholders are included under good governance, and the SAFA framework emphasizes that good governance is necessary to achieve any of the other dimensions of sustainable food and agriculture systems.

International nongovernmental organizations (INGOs) have developed an array of standards to track social sustainability specifically in food systems or more broadly. For example, the International Organization for Standardization launched ISO26000, an updated guide for social responsibility, in 2010. This includes the following seven principles of social responsibility: accountability, transparency, ethical behavior, respect for stakeholder interests, respect for the rule of law, respect for international norms, and respect for human rights. Each principle is clarified and further guidance given on "core subjects" of organizational governance, human rights, labor practices, the environment, fair operating practices, consumer issues, and community involvement.[6] Another example, the Global Reporting Initiative (GRI), is a multistakeholder initiative started in 1997 by the Coalition for Environmentally Responsible Economies (CERES). It became independent in 2002 and now works in collaboration with the UN Environment Programme and the UN Global Compact. GRI's mission is to develop and disseminate its Sustainability Reporting Guidelines, which include in the "Social Standards" category labor practices and decent work, human rights, society (impacts that an organization has on society and local communities), and product responsibility (aspects such as customer health and safety, customer privacy, and labeling). Another nongovernmental organization, Social Accountability International, works more independently to advance human rights in the workplace through compliance with its standard for decent work, SA8000, the first auditable social responsibility standard for decent workplaces. It addresses child labor, forced or compulsory labor, health and safety, freedom of association and the right to collective bargaining, discrimination, disciplinary practices, working hours, remuneration, and management systems.[7] The Ethical Trading Initiative and Fair Trade are other prominent international programs that administer standards systems to protect workers' rights.

II. SOCIAL SUSTAINABILITY IN THE UNITED STATES

Social sustainability, with a solid grounding in internationally recognized human rights, clearly is an important part of standards and expectations in other parts of the world and has been clarified and standardized through intergovernmental agencies and INGOs. Yet in the United States, we seem to have decided that social sustainability is not very important to measure and monitor. There are vast amounts of data, going back in some instances to the mid-1800s, related to economic and environmental sustainability of farms and farmers, including yields and sales of individual crops, economic returns, and specific environmental resources such as soils. How can we explain the relative paucity of social sustainability data and indicators? The lack of attention to social sustainability is a political choice;[8] in the United States, it reflects the current U.S. neoliberal socioeconomic system, in which social implications and impacts of food and agricultural policies tend to be dismissed or portrayed as regrettable consequences of unwise individual choices.[9]

Going further back, lack of attention to social sustainability reflects a political turning point in which the U.S. government decided that economic, social, and cultural rights were best met through market forces, rather than public policy. This seemed important during the Cold War to differentiate the United States, with its capitalist economic system, from communist and socialist governments. This decision was behind the United States' signing and ratifying the UN Covenant on Civil and Political Rights in 1966, but not the sister Covenant on Economic, Social and Cultural Rights; almost all other countries have ratified both.[10] What this has meant in practice is that mainly people in the United States who have adequate economic means enjoy economic, social, and cultural rights, including the right to food, that people in most other countries consider to be the state's responsibility to ensure. Only over the last few decades have social inequities and health disparities related to poor food access and diets stemming from poverty and marginalization been on the public agenda. "Food desert," a place without access to healthy food, has become a common term. Farmworkers' dismal working conditions and lack of a living wage are beginning to receive more press coverage, and the plight of other workers in food systems (e.g., restaurant workers, fast-food workers, employees of big-box grocers) has made headlines over the last year. Far from being individual choices, these disparities in social well-being, access to good jobs, and healthy diets are embedded in the U.S. food system. The goods and services that relatively wealthy people pay very little of their income to enjoy depend on continued exploitation of relatively poor and marginalized people. The lack of access to these goods and services, which most of the world considers to be human rights, is a stunning market failure; yet little has been done to compensate for it.

Indicators of social sustainability are not defined with as much agreement as environmental and economic indicators in the United States;[11] the failure to articulate them clearly results in failure to work on social sustainability strategically and to make steady progress. Many people claim that social sustainability is harder to measure than environmental or economic sustainability.[12] It is indeed true that some of the data needed to address common themes in social sustainability within food systems are less available and accessible in the United States, and often not collected at all. For example, reliable longitudinal data over several decades on farmworker health and mortality and accident rates are impossible to get, and some of these data are inconsistent over different time periods, making comparisons between places and determining trends difficult at best.[13] However, social sustainability is not intrinsically more difficult to assess and monitor than other forms of sustainability.

III. HOW IS SOCIAL SUSTAINABILITY DEFINED, PARTICULARLY IN FOOD SYSTEMS?

Social sustainability is relevant in many stages of the development of standards systems, plans, and principles. During initiation, who makes the decision to undertake the project, and how much power do they hold? Who is invited to participate? In setting goals or principles, which aspects of social sustainability are included? Which are measured, and how rigorously? Who measures? In implementing a sustainability plan, who is held accountable and who monitors the plan? Does government or do other funding sources, such as private foundations, listen to the people monitoring the plan and make adjustments? Each of these questions contributes to the ultimate legitimacy of the principles, standards, or plan.

Several authors have tried to compile the most common areas or themes within social sustainability; overlaps and commonalities are obvious from the brief introduction to international work in the earlier section. Murphy reviewed eight bodies of literature and grouped policy-relevant social sustainability themes into equity, awareness for sustainability, participation, and social cohesion.[14] Magis and Shinn traced social sustainability to three major sources of literature and work: human-centered development, sustainability (drawing heavily on UN documents and conferences), and community well-being.[15] Colantonio reviewed social sustainability literature between 1992 and 2006 to find key themes and claimed that a shift has occurred from more "traditional" themes such as equity, poverty reduction, and livelihoods to "softer" and less tangible themes such as happiness, identity, and the value of social networks.[16] Nelson and Tallontire identified a shift in global agrifood value chain standards from those that focus on retailers' and other companies' need to manage risk and ensure global supply to those that place small-scale producers and vulnerable workers in the center as active participants who must have a voice in administering standards.[17] Derkx and Glasbergen documented efforts to bring together leading standards systems in "meta-governance" initiatives.[18] A joint Initiative on Corporate Accountability and Workers' Rights was established in 2003, including the Ethical Trading Initiative and Social Accountability International, primarily focused on textile industry workers. This initiative was concluded in 2007 without fully meeting its goals of replacing the standards of member organizations with a common standard. On the other hand, attempts to harmonize organic agriculture standards have been far more successful;[19] note that these standards do not include standards for social sustainability. Colantonio suggested that the inability to reach agreement on core themes of social sustainability is due to the different worldviews of researchers.[20] This may apply to social standards created by companies and multistakeholder teams as well.

The right to participate in decision making on issues affecting one's well-being is part of the suite of economic, social, and cultural rights and an element of rights-based approaches to development. In food systems, when marginalized and vulnerable people do not participate effectively in decision making, policies and regulations inevitably favor those who already have power. Regarding food security, this means that wealthy people will continue to have better access to adequate amounts of healthy food than poor or otherwise economically and politically marginalized people, and the latter will not get fair shares of wealth and profits in the food system, proportional to the money, time, and labor they invest. Lobbyists and legislators financed by wealthy patrons continue to push for public policies that keep certain groups (primarily people of color and poor people, but also smaller-scale farmers and fishermen) disempowered, poor, and invisible in the U.S. food system.

The United States has no nationwide agreement, nor a process for discussing and reaching agreement, on social sustainability in food systems or a strategy for achieving it. Most of the work on social sustainability has been through development of voluntary standards systems or state, municipal, and regional food plans. These standards systems and food plans, while very different in their intended uses and contributors, are similar in that the most legitimate outcomes are developed through multiactor processes, and they help to operationalize the meaning of social sustainability. Standards systems are generally used by businesses (including farms) to set or demonstrate a quality threshold and provide assurance that the business is fulfilling its customers' expectations in certain domains. It is usually a marketing tool, and many reports and articles are available comparing the quality of different standard-setting processes and their implications for different types of farms or businesses. Applicability of standards across different sizes of farms or businesses is especially contentious, as some issues only emerge on very large or very small scales. The ISEAL Alliance, a global membership organization for sustainability standards, has created Codes of Good Practice and Credibility Standards that strive to provide some consistency across these processes and their outcomes.

The blossoming of private voluntary standards systems as a global phenomenon has received a great deal of scholarly attention. There is agreement that it is associated with the dominance of neoliberal market-oriented "solutions" to social problems,[21] in preference to reliance on public policy (law and regulations). They are also a manifestation of changes in consumer demand, with growing awareness of the "backstory" of many products and poor wages and working conditions in supply chains. Pressure to demonstrate corporate accountability and responsibility in food and agriculture is growing, especially since the figurative and literal distance between producer and consumer has grown. When consumers have no direct contact with producers, they often want a guarantee that the products they are buying are safe and healthy, and that supply chains have practices congruent with their values. In addition, standards systems can help companies to manage risk by increasing their control of quality in global value chains. Standards systems meet all of these demands, and in ways that allow producers to create new niches for themselves in very competitive food markets.

The first U.S. public compilation of food and agricultural standards was the reference library created by volunteers who developed the first Sustainable Agriculture Standard;[22] the author co-chaired the Social Sustainability Sub-Committee for nearly two years at the beginning of its work. Since this library was initiated, the International Trade Center has created a Standards Map that includes 113 agricultural standards in North America. The most common social themes in food and agricultural standards systems were reflected in the LEO-4000 standard: work agreements; wages; benefits; working hours; child labor; voluntary labor; nondiscrimination; freedom of association; protection from violence and harassment; human resource management; healthy and safety; workplace conditions; worker housing; stakeholder and community engagement; and local and regional sourcing, sales, services, and support. In each of these areas, the Standards Committee developed indicators that were optional or required at each of four levels of performance (bronze, silver, gold, and platinum).

LEO-4000 encountered tremendous opposition from different quarters. The very notion of developing a standards system for sustainable agriculture created rifts among organic and sustainable agriculture supporters. While resistance to this standard had many contributing factors, some of the strongest seemed to be fears that the standards would lead to tighter regulations; ignorance of the actual content of the standard and its development process (for example, although continuous improvement was part of the standard from the beginning, some people objected that standards for sustainable agriculture could not be developed because it is a constantly moving target); fears of potential competition with organic agriculture; and lack of awareness that similar standards were being and had already been created in the European Union and by many transnational companies that control food supply chains. That is, the private voluntary standards train had long left the station, and it was not being driven by organic or sustainable agriculture supporters. The only way that organic and sustainable agriculture supporters could hope to jump on that train was to participate in a fair multistakeholder process—not disparage it and hope it would go away. A completely different set of objections came from agribusiness and lobbyists for commodity groups, who thought the process was illegitimate because organic and environmental interests had too much influence.

Social sustainability is part of several other U.S. food system standards systems that have been developed through multistakeholder teams that often include representatives from farmer and labor organizations, notably the Agricultural Justice Project's Food Justice standards and the Equitable Table Initiative. The Domestic Fair Trade Association evaluates different U.S. social sustainability certifications for congruence with principles they have established. The first social sustainability certification in U.S. food systems was based on the Food Justice standard, which includes workers' rights to freedom of association and collective bargaining, fair wages and benefits for workers, fair and equitable contracts for farmers and buyers, fair pricing for farmers, clear conflict resolution policies for farmers or food business owners/managers and workers, the rights of indigenous peoples, workplace health and safety, farmworker housing, interns and apprentices, and children on farms. Other U.S. standards systems that include social sustainability have been developed for particular crops, such as the Stewardship Index for Specialty Crops; the Lodi Rules (wine grapes); and Protected Harvest (now administering the Lodi Rules, but also certifying stone fruit, citrus, and mushrooms).

In addition, almost every large company involved in a food system activity, from input production through retailing and waste management, produces a corporate social responsibility report that includes social sustainability. Walmart, the highest-volume grocery retailer in the United States, includes the following within its "social responsibility" theme: ethical sourcing, global audits, global women's economic empowerment, hunger relief, healthier food, giving, and disaster relief. While all standards systems include indicators of some kind, their significance, transparency, and certification methods vary considerably. Walmart's Sustainability Index and track record have been critiqued heavily,[23] although many business commentators consider its effort to improve the sustainability of its supply chains to be significant and laudable. It is safe to say that some aspects of social sustainability are likely to get little attention from a company that specializes in low prices, such as Walmart.[24] Livable wages; transparency regarding practices and the flow of money; and more equitable sharing of profits among shareholders, managers, and workers in global supply chains are sticking points for social sustainability in a capitalist economy: few companies prioritize the well-being of their employees over higher profits to shareholders or owners. The public knows little about what is actually happening within most companies, behind glossy corporate responsibility reports and Web sites. Playing its "watchdog" role, civil society has revealed problems with specific bad actors, but only on an ad hoc basis.

IV. SOCIAL SUSTAINABILITY IN U.S. FOOD SYSTEM PRINCIPLES AND FOOD PLANS

Two other places where action is taking place to define and operationalize social sustainability in the U.S. food system are through principles developed by multiactor teams and in state, regional, and municipal food plans. The best example of the former is the Principles of a Healthy, Sustainable Food System, developed in 2010 by the American Public Health Association, the American Nursing Association, the Academy of Nutrition and Dietetics (formerly the American Dietetic Association), and the American Planning Association. Social sustainability appears throughout these principles: for example, the specifications that healthy, sustainable food systems are culturally diverse and support fair and just communities for everyone. This set of principles is particularly interesting because social and economic aspirations are interwoven in the goals to be health-promoting, sustainable, resilient, diverse, fair, economically balanced, and transparent. While the principles echo many economic, social, and cultural rights, they are not presented as basic human rights but rather as aspirations. Good governance is not included in this set of principles.

A recently articulated set of principles for a "just, equitable and sustainable food system" came from a group supported by the W. K. Kellogg Foundation and hosted by the Union of Concerned Scientists. These principles are noteworthy for emphasizing the importance of leadership, decision making, and community self-determination by small- and medium-scale farmers, farmers of color, farmworkers, food chain workers, indigenous people, low-income people, women, and communities of color. They focus on policy changes needed to create a just, equitable, and sustainable food system.[25]

In the vacuum created by the lack of any food system strategy at the federal level, states, municipalities, and regional agencies and organizations have created their own plans, often beginning with a food assessment to identify problems and assets in the geographic area of concern. Establishment of principles may follow food assessment, although the order in which organizations work on different elements of a strategy varies tremendously. Some develop principles, goals, and indicators before completing a food assessment to find out the current status of the food system in their area; some never go beyond the assessment to create a strategy for change. Clearly expressed social sustainability principles are part of much of this new work on food system strategy. Sometimes these principles appear as charters, such as the Michigan Good Food Charter, the Minnesota Food Charter, and the Philadelphia Food Charter. Charters usually are simply statements of values, principles, or operating methods, but charters or statements of principles can be the basis of full-blown strategies for achieving a different food future.

Food security, usually expressed as improved access to healthy food by all people, is a component of food charters, vision, strategies, and plans at every scale when they are products of multiactor processes. This may be one of the most compelling reasons to include different perspectives in food system planning: including voices of marginalized people ensures that strategies will accommodate the needs of people who are not well served by the current U.S. food system. In New England, a recently published vision for the future of food and farming in 2060[26] included "healthy food for all" as one of the key principles and recognized that explicit public policy to broaden access to healthy food is necessary to meet this principle. Although the vision was coauthored by a small group, it was vetted twice in large regional New England food summits, and comments of people from many different parts of the food system were incorporated. This vision was unusual in introducing the right to food and the right to work as important components of social sustainability.

The author was part of the process of developing a Maine Food Strategy, and the group set "healthy and affordable food for all Maine residents" as one of its five goals (but not stated as a basic human right). Among state food system strategies, Vermont's is outstanding for its clear goals, strategies, indicators, monitoring, and communications. The twenty-five goals in Vermont's Farm to Plate Initiative include food security: "all Vermonters will have access to fresh, nutritionally balanced food that they can afford" and the following are strategies to implement this goal: (1) increase the ability to integrate local purchasing into supplemental assistance programs (e.g., fruits and vegetables vouchers, Farm to Family coupons); (2) perform impact evaluations of organizations and programs that focus on increasing food access; (3) share best practices across all food security stakeholder groups; (4) establish and fully fund gleaning programs and coordinators in every region of the state by 2014; and (5) identify and address the needs of food insecure groups that are unserved or underserved (i.e., immigrants, elders, and the homeless).[27] In contrast, relatively few state plans or strategies include clear indicators and mechanisms for monitoring change over time. This often reflects lack of state resources to collect and report on data consistently, or lack of government investment in the planning process at all. Plans led by nongovernmental organizations face especially tight capacity restrictions on collecting data, monitoring trends, and issuing consistent reports over sufficient time to detect improvements.

The substantial work happening across municipalities and states to develop strategies to improve the sustainability of their food systems would be helped significantly by better data. However, there is no official link between the Agricultural Census and state planning processes; and each state is acting on its own accord and using whatever data it can find. The U.S. Department of Agriculture's Food Environment Atlas has been very useful to many state groups; but when desired data have not been collected, or not at the scale that planners need, there are no clear ways to register requests for these additional data to be collected. Other commonalities among municipal, state, and regional food plans that may weaken them are, first, that they seldom use the language of human rights, such as the right to food. This language could connect food charters and plans with internationally agreed-upon rights and standards based on them, with a deep history of work devoted to best ways to realize them. Second, these plans tend to ignore the need for continued multiactor participation in monitoring and evaluating plans and strategies, even when the plan was created through a multiactor process.

V. CONNECTIONS BETWEEN SOCIAL SUSTAINABILITY AND AGRARIANISM

Agrarianism is a sociopolitical philosophy that values farming as an incubator of important virtues and rural farm-based life as superior to city life. U.S.

agrarian writing has its roots in Roman thought (such as Virgil's *Georgics*) and English and European philosophers. Thomas Jefferson was an early advocate of agrarianism. He wrote in a letter to John Jay in 1785 that, "Cultivators of the earth are the most valuable citizens. They are the most vigorous, the most independent, the most virtuous, & they are tied to their country & wedded to its liberty & interests by the most lasting bonds."[28]

Agrarianism has surged and retreated during U.S. history, with a renaissance in the 1940s and more recently during "back-to-the-land" movements influenced by writers such as Aldo Leopold, Victor Davis Hanson, Helen and Scott Nearing, Wes Jackson, and Wendell Berry.[29]

Themes that pervade agrarian writing are the inherent dignity of farming, which "city people" often fail to understand in their distraction-fragmented lives. Cultivation of the soil and direct contact with nature give the farmer wisdom, knowledge of seasonal cycles, and the rhythm of birth and death— all essential to moral integrity and mental and physical health. In addition, the hard work of farming from dawn to dusk builds character and self-discipline. Agrarians write about attachment to place and intimate knowledge of that place providing a sense of tradition and belonging to a community. These are considered to be important to psychological health; the agricultural community often is idealized in agrarian writing as a place of order, cooperation, and social harmony. The self-reliance and independence of farmers are seen to be essential linchpins of democracy.[30] Of course, analysts of agrarian philosophy do not accept these themes as "truths" and there is a growing body of critical writing on agrarianism;[31] but they are still important components of widespread U.S. perceptions of farming and rural life.

Seen from the perspective of human rights and social justice, agrarianism has many gaps. It is deeply patriarchal; women and the earth are subservient to male-imposed order and control, and there is almost no discussion of the exploitation of women and children.[32] Child labor is perceived positively; children learn from their elders by working under their tutelage, as in traditional societies. Agrarian writing often ignores the patterns of racial exploitation within the U.S. food system. For example, it has very little to say about the original theft of U.S. farmland from Native Americans or about slavery as a social evil (with a few notable exceptions, including one of the earliest U.S. agrarians, J. Hector St. John de Crèvecoeur, in his 1782 *Letters from an American Farmer*[33]). While more equitable redistribution of land is a strong tenet of agrarianism in many other countries, U.S. agrarian writers tend to glorify private property. The need for fair prices for farmers is a steady undercurrent in agrarian writing, but food insecurity and lack of equitable access to healthy food seldom are not abiding concerns. Similarly, there is little deep analysis of structural barriers that poor consumers or small-scale farmers encounter.

Current U.S. values with respect to agriculture and farmers draw heavily on agrarianism. The emphasis on the farmer, particularly the small-scale farmer, as the most important element of the food system comes directly from agrarianism. Farmers are sometimes portrayed as larger than life (e.g., Communities in Support of Agriculture in Western Massachusetts has a "Local Hero" campaign to recognize farmers). As late as the mid-1900s, small-scale farmers were perceived to be backward hicks; but they have moved closer to the "noble yeoman" archetype in the public mind over the last couple of decades. The iconography of farming displayed on packaged food to attract customers emphasizes bucolic images of nature, rather than the reality of mucking out stalls and long dusty hours on a tractor. While many people in the United States have little understanding of farming life or the economic realities of farming, they prefer small-scale farms and believe that corporations and private money control most food system decisions. The strong opposition to industrialized farming and the negative perception of agribusiness are related to agrarian ideas. Farmers' markets, where customers can talk with farmers (or at least people they hire to work for them), have increased rapidly in numbers; and community-supported agriculture also is growing. This is happening despite the simultaneous trend to eat more food away from home and more prepackaged-for-convenience food; restaurants in large cities are capitalizing on both trends by sourcing fresh, local food.

Agrarian themes and ideas have permeated U.S. perceptions of agriculture and social sustainability within food systems, casting white male farmers in the limelight. The gaps in agrarianism may help to explain why many other aspects of social sustainability—wages and working conditions for farmworkers, discrimination against indigenous people, gender discrimination and exploitation, food system governance, land loss by people of color—have gotten short shrift in the United States. The emphasis in agrarianism on farmers supporting society leads easily into the ideology that U.S. farmers have a responsibility to "feed the world," and away from the idea that people feeding themselves is the best strategy for addressing world hunger. The glorification of (white, male) farmers as independent, self-reliant pillars of democracy leaves little space for multiactor governance mechanisms that include marginalized people.

VI. CONCLUSION

Economic, social, and cultural rights including the right to food are increasingly incorporated into the framework for social sustainability at the international level. The role of governance, including participation of all relevant stakeholders in food system decisions, is also moving to greater prominence. U.S. food principles and food strategies are not keeping up with these expanding perceptions of the meaning of social sustainability in food systems.

Meanwhile, private voluntary standards and quality standards imposed on suppliers by corporations are moving into the lacuna created by inattention at the federal level to food system sustainability and how it can be operationalized. Yet private-sector corporate social responsibility is far less powerful as a means of implementing and making progress toward social sustainability than law or regulations; there are few mechanisms to hold corporations accountable for the well-being of their customers within national and international law. Private-sector standards also enact multiple contradictions and perpetuate an audit culture,[34] rather than helping to construct a social and solidarity economy in which social sustainability becomes the norm.

Rights-based indicators (e.g., violations of the right to food) are especially promising as ways to monitor social sustainability because they include process as well as outcomes. Therefore, they show whether people are able to have effective political participation in making decisions about how food systems are governed, what they produce, and at what costs. Without such participation by people who are not currently food secure, there is insufficient incentive to change current practices and policies and insufficient knowledge to make changes in ways that will bring long-term food security.

NOTES

1. International Labor Organization (ILO), *ILO Declaration on Social Justice for a Fair Globalization.* Adopted by the International Labor Conference at its 97th Session, Geneva, Switzerland, June 10, 2008.

2. HLPE, *Investing in Smallholder Agriculture for Food Security: A report by the High Level Panel of Experts on Food Security and Nutrition of the Committee on World Food Security,* Rome (2013); Beverly McIntyre, Hans Herren, Judi Wakhungu, and Robert Watson (Eds.), *Agriculture at the Crossroads: International Assessment of Agricultural Knowledge, Science & Technology for Development.* Global Report (2009).

3. Food and Agriculture Organization (FAO), *About FAO* (2014), http://www.fao.org/about/en/.

4. Food and Agriculture Organization (FAO), *Short History of SAFA* (2014), http://www.fao.org/fileadmin/templates/nr/sustainability_pathways/docs/SAFA_History.pdf.

5. Food and Agriculture Organization (FAO), *Sustainable Food and Agriculture Assessment Indicators* (2013), http://www.fao.org/fileadmin/templates/nr/sustainability_pathways/docs/SAFA_Indicators_final_19122013.pdf.

6. International Organization for Standards (ISO), ISO26000:2010, https://www.iso.org/obp/ui/#iso:std:iso:26000:ed-1:v1:en.

7. Social Accountability International (SAI), SA8000:2014, http://www.sa-intl.org/index.cfm?fuseaction=Page.ViewPage&pageId=1689.

8. Andrea Colantonio, Social Sustainability: Linking Research to Policy and Practice (2009), *available at* http://www.lse.ac.uk/researchandexpertise/experts/profile.aspx?KeyValue=a.colantonio%40lse.ac.uk.

9. Marion Nestle, Food Politics: How the Food Industry Influences Nutrition and Health (2007).

10. See https://treaties.un.org/Pages/Treaties.aspx?id=4&subid=A&lang=en for signatures and ratifications of both human rights covenants.

11. Kevin Murphy, *The Social Pillar of Sustainable Development: A Literature Review and Framework for Policy Analysis*, 8 Sustain.: Sci., Practice & Pol'y 15 (2012).

12. Magnus Boström, *A Missing Pillar? Challenges in Theorizing and Practicing Social Sustainability: Introduction to the Special Issue*, 8 Sustain.: Sci., Practice & Pol'y 3 (2012).

13. Molly D. Anderson, *Charting Growth to Good Food: Developing Indicators and Measures of Good Food* (2009), http://www.wallacecenter.org/resourcelibrary/charting-growth-report.html.

14. Murphy, *supra* note 11.

15. Kristen Magis and Craig Shinn, *Emergent Principles of Social Sustainability*, *in* Understanding the Social Dimension of Sustainability, 15–44 (J. Dillard, V. Dujon, and M. King, eds., 2009).

16. Colantonio, *supra* note 8.

17. Valerie Nelson and Anne Tallontire, *Battlefields of Ideas: Changing Narratives and Power Dynamics in Private Standards in Global Agricultural Value Chains*, 31 Agric. & Hum. Values 481 (2014).

18. Boudewijn Derkx and Pieter Glasbergen, *Elaborating Global Private Meta-Governance: An Inventory in the Realm of Voluntary Sustainability Standards*, 27 Global Envtl. Change 41 (2014).

19. *Id.*

20. Colantonio, *supra* note 8.

21. Lawrence Busch, *Governance in the Age of Global Markets: Challenges, Limits and Consequences*, 31 Agric. & Hum. Values 513 (2014).

22. The reference library for LEO-4000 is available at https://sites.google.com/a/leonardoacademy.org/sustainableag-referencelibrary/standards and includes the text of 89 standards relevant to food systems. The LEO-4000 draft standard (dated November 13,

2015) is available at http://leonardoacademy.org/services/standards/agstandard.html. See http://www.standardsmap.org/identify for standards included in the International Trade Centre database.

23. See, e.g., Institute for Local Self-Reliance (ILSR), *Top 10 Ways Walmart Fails on Sustainability* (2012), http://www.ilsr.org/wp-content/uploads/2012/04/topten-walmart sustainability.pdf.

24. Susie Cagle, *Walmart Wants to Be Sustainable? It Should Start with Its Labor Force* (2012), http://grist.org/news/walmart-wants-to-be-sustainable-it-should-start-with-its-labor-force/.

25. Amelia Moore, Union of Concerned Scientists, *Good Food for All, Working Definition of a Good Food System*, personal communication, August 28, 2014.

26. Brian Donahue, Joanne Burke, Molly Anderson, Amanda Beal, Tom Kelly, Mark Lapping, Hannah Ramer, Russell Libby, and Linda Berlin, *A New England Food Vision* (2014), http:// www.foodsolutionsne.org/sites/default/files/LowResNEFV_0.pdf.

27. See http://mainefoodstrategy.org/ for more information on Maine's progress, and http://www.vtfarmtoplate.com/getting-to-2020 for Goals and Indicators of Vermont's state plan.

28. Thomas Jefferson, Letter to John Jay, August 23, 1785, *available at* http://en.wikisource .org/wiki/Letter_to_John_Jay_-_August_23,_1785.

29. Edwin C. Hagenstein, Sara M. Gregg, and Brian Donahue, eds., American Georgics: Writings on Farming, Culture and the Land (2011).

30. Allan Carlson, The New Agrarian Mind: The Movement toward Decentralist Thought in Twentieth-Century America (2007); Eric T. Freyfogle, Agrarianism and the Good Society. Land, Culture, Conflict and Hope (2007); Hagenstein et al., *supra* note 29; Paul B. Thompson, The Agrarian Vision: Sustainability and Environmental Ethics (2010).

31. Liz Carlisle, *Critical Agrarianism*, 29 Renewable Agric. & Food Sys. 135 (2014).

32. Linda J. Borish, *"Another Domestic Beast of Burden": New England Farm Women's Work and Well-Being in the 19th Century*, 18 J. Am. Culture (1993); Rebecca Sharpless, *Southern Women and the Land*, 67 Agric. Hist. 30 (1993).

33. J. Hector St. John de Crèvecoeur, *Letter IX*. Letters from an American Farmer (1782) *available at* http://avalon.law.yale.edu/18th_century/letter_09.asp.

34. Busch, *supra* note 21, at 513–23.

II: Views from Within
the Food System
THE FARMER, THE CONSUMER, AND THE WORKER

5 Community Agriculture and the Undoing of Industrial Culture

Josh Slotnick, University of Montana

It's early. The bustle has yet to begin here in the sit-down deli section of a giant gorgeous natural food grocery store in Missoula, Montana. In another hour or so, people across the age spectrum will fill the place, stuffing shopping carts, getting a Sunday morning coffee, or taking advantage of the lush breakfast burrito bar, before heading up to the ski hill. High ceilings, tasteful lighting, exposed ductwork, and lots of wood all from a suite of "natural" color values; this scene exemplifies the aesthetic, and, more importantly, the bounty, of the best food in America. This wonderful independent store in a small city in the West fills its shelves with organic products, the produce section boasts an allegiance to local, and even the buffed-out salad bar knows its farmers. This same town also sports two large farmers' markets, half a dozen community-supported agriculture organizations (CSAs), a farm-to-school program, a farm-to-college program, and some fantastic nonprofits that blend food security with advocacy, land preservation, and social justice. Twenty years ago, all of these elements of change would have been impossibilities, if not pieces of an impossible dream. Yet down the road from here, both north and south, two separate Walmarts pump out food and food-like products at a dizzying mechanistic pace.

I have been farming in the Missoula Valley for the last twenty-odd years, selling locally, and watching carefully the evolution of our food system. I have the sense that the changes in our food system have been as much in degree as they have been in kind; it is as if the volume on the food system as a whole has been cranked up to 11. We used to have a few Safeways and an Albertson's; now those two grocery stores have merged. We have a pair of Walmarts as well, one as big as an airport. Our cozy little natural food shop grew into this beautiful full-service grocery store; our one quiet farmers' market has become two bustling markets, and we have added the above list of alternative accoutrements to the modern food system as well (CSAs, advocacy organizations, farm-to-school, etc.). However, the situation begs some questions. Given the tremendous effort made by a lot of hard-working, organized, and razor-sharp folks across the country over the last two and one-half decades, why have the changes in our food system not cut deeper into, or even altered the pace of growth of, the industrial food system? Why have we not been more successful, relatively speaking? And what then needs to be done? The following is one farmer's take on all that, and a pitch for us to grow food together, Community Agriculture, wherever we can, for more than the sake of sustenance.

I. HUMAN-SCALE FOOD

Smart and strategic people have done jujitsu with local politics as well as social trends, legislating land preservation, allocating funds for school lunch programs, and creating civic agriculture projects that often speak to the best of our natures, but these elements of second-decade twenty-first–century America remain on the margins. For the most part, the mainstream of the food system has resisted humanization. The mainstream of our food system works on an industrial scale, not a human one.

Food grown, distributed, and finally prepared and shared on a human scale embodies many of the traits you would expect: its production relies on empowered people as well as appropriate machines, and it depends on a rich understanding of specific ecologies. The distribution networks are relatively tight and easy to grasp intellectually—people buy food mostly whole, only a few steps removed from where it began, and then add their own labor in the creation of meals. All of those characteristics reflect an insertion of care and personal investment, not just effort, at every step from the farm to the table. As anonymous as is industrial food, human-scale food is intimate—someone cared and put a bit of their best selves into making a meal happen. This level of intimacy creates a knowable and understandable story. When we sit down to eat, the table pulls together well-made food and all its history—where it came from, how it was grown, and who grew it, as well as how it was chopped and cooked. In the best scenario all those steps should be knowable, not because understanding moves us closer to finding purity in our food or because knowing the history of our food facilitates ever more indulgence in personal health, but because active caring about our food requires knowledge. If we want the treatment of farmland and farm people to shine back to us our most humane values, then we need to know the history of our food (knowledge). We can then begin to create a system where intimacy can supplant anonymity whenever possible.

As for the nature of those humane values, I believe these are the sorts of ethical tenets generally understood in elementary schools everywhere: compassion, justice, and fairness. Put into action, this means the people involved in growing food should enjoy levels of autonomy typically reserved for the middle class; the work—farming—should call out for their care, not just their backs, as such work is craft, not manufacturing. Agricultural communities should experience prosperity enough to maintain vibrant cultures, and these types of communities should be well dispersed across the landscape. Good farming requires a partnership with living land; consequently, we must treat the ground as we would husband a loved animal. Scaled up, all of that sounds pipe-dreamy, if not ridiculous, to modern ears, right? A paean to a cozy Jeffersonian world that never existed, or a Wendell Berry play, exquisitely wrought, but certainly not possible. To keep digging in here begs yet another question: why does the dream of a human-scale food system, spread deep into the mainstream, seem so far-fetched? The answer, I believe, has not so much to do with food and farming in the specific, but everything to do with our culture at large.

We have an industrial food system because we have an industrial culture. The industrial food system has successfully resisted large-scale, justice-oriented changes, because those changes do not fit smoothly in an industrial culture. I realize those are bold claims, so, let's take a quick look at a few characteristics of an industrial system—a large-scale manufacturing process—and see if the shoe fits. Do we actually have an industrial culture?

II. SEGMENTATION, PRODUCTION, AND PLACELESSNESS: CHARACTERISTICS OF AN INDUSTRIAL SYSTEM

An industrial manufacturing system orients itself toward production; other concerns fall in line to serve this goal. Since Henry Ford, industrial engineers have understood the production gains available from the segmentation of process. The assembly line dumbs down the construction of an object, simplifies a worker's role, and fantastically increases efficiency. I once worked on a fish processing line; I performed my task and passed the fish on, as fast as I could. I had only a limited sense of what happened before my step, or what was to occur later on down the line. This segmentation effectively separated me from a holistic sense of the process. No value or ethical consideration (beyond production) integrated the system. Each step felt stand-alone, under an umbrella of deafening machine noise and fish stink. Given the goal of the endeavor (production), segmentation and separation made great sense. The line moved blazingly fast; fish became meat, by the ton, every hour.

A successful modern manufacturing system evaluates even primary considerations, like place, along a production rationale, as does manufacturing's brother, distribution. The production orientation measure of a place is the commercial essence of globalization; production occurs where production costs are cheapest and sales happen where wholesale and (then) retail opportunities are the most lucrative, regardless of geography. And the whole world is in play. A modern manufacturing system exists apart from the biological and cultural parameters of place; it is placeless, all locations being equal, outside of the economic implications for production. Witness the drift of auto parts manufacturing from the American Midwest to Mexico post-NAFTA, the death of the U.S. steel business in Pennsylvania and the rise of Chinese foundries, or the shift of textile mills from the southeastern United States to Bangladesh and Guatemala. These manufacturing entities moved relatively smoothly to radically different places with only minor temporary hiccups in product availability. The actual specific place of these factories did not matter (does not matter) per se. Steel bolts, distributor caps, T-shirts, were ubiquitous and cheap; they are right now, and seemingly will be for evermore. From a production orientation, placelessness kicks butt. The same sorts of developments fuel the sales and distribution born of this system.

Big-box retailers dominate the market for transactional commerce, and these stores, much like their factory brethren, choose locations along the narrow parameters of production economics. Every American city of a certain size sports a valley built of recognizable architecture and familiar store names. These commercial outlets repeat themselves in name and number across the landscape. If population growth in an area warrants it, the mega-stores arrive, and if the population numbers recede, the mega-stores too disappear. These retailers exist outside the biological or cultural parameters of place, and their proliferation upon the modern world physically spreads the doctrine of placelessness. The Costco Valley here in Missoula looks a lot like the arrangements in Nashville, Saratoga Springs, or Fresno. The homogenizing force of this type of development erodes the culturally prickly as well as the locally fabulous elements from our cities and towns and effectively smooths our country into one bland place. But so what? From an industrial perspective (production), homogenization is a sensible goal, as sameness facilitates efficiency and can only further benefit cost-saving measures (the structure of a Barnes & Noble parking lot anywhere neatly accommodates a Pier 1, no matter the seasonal inclinations of the land, the accents of the people, or the regional cuisine). These types of stores sell cheap stuff, affordable to almost everyone, and isn't that kind of democratic? Furthermore, the mega-stores create jobs. All of that is hard to beat on a purely production metric.

Efficient manufacturing, distribution, and retail systems rely on not just cheap, but replaceable labor, and extract every last drop of productive energy from their employees. Brilliant engineers have designed skill out of work as much as possible, replacing craft with force (visualize the assembly line of large-scale meat cutting or the plating of cheap restaurant food). The steps necessary to production link together neatly, but do not require rich personal investment or the attentive nuance associated with actually making something. The people working in this type of system can be swapped out with new people, nearly seamlessly; consequently, those doing the work feel the production pressure intensely and can literally smell their own expendability (think farm field labor piecework, Amazon's fulfillment center runners on stopwatches, or even car salespeople scrambling to meet sales quotas). These workers must bust it out because if they cannot, there is someone else who will. But the anonymity inherent in the design of these workplaces makes them feel hostile and inhumane. No one will ever become sentimentally attached to these places.

We thrive, and are actually more productive, when we recognize our contributions; we need to belong to others (have community) and enjoy enough relative security to apply our full focus to the job at hand. As I am sure you, dear reader, have experienced, real success requires a high degree of personal investment. Yet we have designed work scenarios (and not just factories) that devalue people, remind them of their eminent replaceability, their lack of specific individual value, and offer little opportunity for community. These production systems ask little of us as people—we are soulless, faceless, and anonymous in the mega-store, the fry line, or the fulfillment warehouse. Hopefully, you are hearing the rhyme of anonymity with the previous paragraphs about placelessness. In an industrial system, neither specific places nor specific people matter much.

III. DOES THE INDUSTRIAL SHOE FIT THE FOOT OF MODERN CULTURE?

It may seem out of place to describe sanitized twenty-first–century American life as "industrial," a word that may inspire Dickensian coal-smoke visions of nineteenth-century British factory-generated misery; but I use this word because it connotes a system where ultimate value rests with the results of production. I believe we have applied some of the essential elements of industrialism—segmentation, separation, and placelessness—to our lives, all in service of the number one industrial goal, production.

In order to, at the least, pay the bills to economically persist, we work our butts off at a steadily increasing rate. We work more than ever before and punctuate our relentless workweeks with time carved out for friends, family, exercise, recreation, diversion, each activity often in its own box, and its own place along an assembly line. This compartmentalization stems from necessity, and the arrangement works, according to our standards, more or less well. These elements arrayed properly, with fortitude, and the good fortune of being born into a degree of privilege, can produce desired middle-class results.

When bumping into my middle-class, middle-aged peers, if the oft-asked question, "how's it going?" is answered honestly, the typical reply sounds something like "swamped, so busy, barely hanging on right now." We expect this of one another, as we expect that we will answer e-mail even as it continues to pour in, like a person shoveling snow in a blizzard or bailing out the ocean with a spoon while the rain falls, all while attending multiple meetings and doing the work we promised to do. This production ethic extends to modern middle-class parenting as well, the relentless overscheduling easing only when the kids drive themselves from one thing to the next (we train them to become us). Our emphasis, when evaluating our lives, appears to be on production results—did you do what needed to get done (meet production goals)? Like the evaluation of any production system, our focus is more on the results of our efforts than the character of the process.

Meeting these intense production expectations requires compartmentalization (the separation of one part of life from another) and a linked system within our own lives. The linked system certainly facilitates production, but reduces the potential for a holistic understanding, or the integration of ethics, and even limits the need for deep personal investment in our own lives (others do for us, so we can spend our time on production). Often the evidence of our fealty to production turns up outside of our jobs: we hit the drive-through for lack of time;

pilot big, three-quarter-empty cars to work; slurp coffee in traffic; and pick up a pizza on the way home. Though we might not be proud of it, we do these things in order to facilitate, again, doing what needs to be done. The production expectations we exact upon one another leave little room for expectations around intention—the character of process of how we live. We have designed work systems with high production expectations, and we respond accordingly. Those active responses demand most of our energy, and the necessity of production devalues nearly everything else. How can we meet these production expectations and have the mental and physical energy, or the time in the day, to exercise value-driven intention around anything that does not further our production efforts? Put simply, most of us cannot.

We have encouraged our colleges and universities to train people for gainful employment. Students declare a major, and their families (who save, borrow, skimp, and document their way to paying for the schooling) sensibly ask about the return on such investment, "What are you going to do with that major?" The underlying assumption in this question echoes another version of placelessness—people should make themselves marketable and then go where the job opportunities are most lucrative, all locations being relatively equal beyond economic implications, and the whole country is in play. As a people, we have now indeed spread across the landscape like dry leaves, when most everyone you meet is from somewhere else. Those able to make themselves marketable can financially reap the rewards of placelessness, moving to a new gig inside the Beltway, Silicon Valley, or the I-5 corridor. People born into cultures of lower expectations, however, often stay closer to where they began, but reel even more intensely under production pressures than their middle-class counterparts. If you earn crappy wages, the way to make your arrangement pay is through volume—that is, work more.

Here in Montana we like working so much we regularly lead the country in people with multiple jobs, all so we can earn a median income well below the national average. The texture of low-wage work (fast food, assembly lines, fulfillment centers, piecework in farm fields) only reflects part of the issue. The physical intensity, low wages, and time commitments—shift work with no flexibility—associated with this work make intentionality outside of work even more difficult than for folks in the middle class. Put more bluntly, and more obviously, working-class people have even less room in their lives than middle-class people to exercise intentionality. (See Tracie McMillan's wonderful book, *The American Way of Eating*.)

So, it seems for the middle class and the poor alike, we have aligned our lives around production, and we have designed systems within our culture to facilitate efficiency. By this description, we have an industrial culture, and an industrial culture requires industrial food. I hope to convince you that we have the food system we deserve.

IV. THE LOGIC OF INDUSTRIAL FOOD

Industrial food reflects the logic of an industrial culture; it is placeless, the origins can barely be known. For example, the shrink-wrapped convenience store muffin is a product of national assemblage; it is anonymous, hailing from nowhere in particular, instead finding its genesis in a whole slew of disconnected places, all far away. Assembling this muffin required a segmented process demanding little personally of the people involved, and its eventual consumer is equally disconnected from its history. All that being said, I believe industrial food literally feeds an industrial culture, not because they are both cut from the same bolt of cloth, but because an industrial culture leaves little room for anything else. Production, the central value of industry, is ultimately a concern with results, not with process (how you get where you are going does not matter so much as that you make it on time). Intention, how we deliberately choose to do something, reflects an entirely different value: how you get where you are going matters as much as arriving. As described above, our culture exacts tremendously high production expectations of us, and these expectations reveal themselves in the intensity of modern life. Meeting our production goals requires most all of our available time and energy. In the modern world, industrial food facilitates meeting production goals. Because it is quick, cheap, and ubiquitous, industrial food demands little time and energy of us, leaving more of those resources available in the service of production. We cannot reasonably demand intention around food—a real meal made from whole, known, human-scale food, shared with our household—unless we ratchet back our expectations around production. Our industrially segmented, production-oriented lives, regardless of class, require readily available, low-cost, low-labor food. Our culture was not made for the microwave, the superstore, and the drive-through; those ubiquitous elements of life persist and multiply because they fit our culture. Put more simply, the drive-through helps you get your work done, as does the microwave and the superstore. Human-scale food sounds fine and dandy, but there is just not room for it, on scale, unless we become a nation of the retired and well-to-do.

Food and farming activists and academics have attempted to chew away at the edges of the industrial food system by making human-scale food more available, and we have enjoyed some great successes. More farmers' markets and more CSAs serve more people than ever, and we have better grocery stores than twenty years ago. We have effectively made human-scale food more ubiquitous, cheap, and quick than it was twenty years ago, but because of the personal investment required of human-scale food, we will never really compete on these measures in an industrial context. If we want to see wholesale cultural change, we need a new context, a swapping of evaluative metrics from ubiquitous, cheap, and quick in the service of production, to the values associated with human-scale food (care actively expressed for people and place). In order to change the culture toward those values, we must lead lives where intention—how we live—is as important as how much we get done, and therein lies the entry point for community agriculture.

V. COMMUNITY AGRICULTURE

Over the last decade or so, community agriculture projects have popped up in major cities and smaller communities across the country. Detroit's much publicized literal blossoming from the ruins of dead neighborhoods and defunct industrial parks to a city of farms and gardens leads a charge echoed from Los Angeles to New Orleans to St. Louis to Hana, Hawaii. "Urban agriculture" is often used to describe these efforts, and that term often accurately reflects the geography of the work, but not its effect. I believe most of this work is more aptly termed "Community Agriculture." To define Community Agriculture (CA) I will begin with what it is not; it is not community gardening, where people rent plots for home gardens off-site from their actual households. In Community Agriculture people do not divvy up a chunk of ground for individual gardens; rather they manage a piece of land as an entirety. The work done on these farms and gardens is not done by professional farmworkers solely, but by people who will benefit from the experience (they may or may not be employed there); these people could be adult volunteers, paid youth in job training or therapy, or school kids at camp—really anyone. The food these farms and gardens grow is typically not sold into the industrial economy, where it would become storyless, anonymous, and well traveled. If CA growers do sell their food, they wave their flag high, celebrating, rather than burying the creation story of their food. That means selling at CSAs, farmers' markets, farm stands, or local stores, where the food's history is overtly identified if not celebrated. Typically, CA organizations give away some and sometimes all of their food to those who need it, using existing distribution infrastructure like food pantries and homeless shelters, or doing their own distribution, and workers and volunteers share the produce as well. Sales from CA generate funds and outreach support to further the effort, rather than to maintain the economic livelihood of one family or business. Given the parameters so described, you probably would guess that the goals of a typical CA operation differ from standard agriculture (industrial or human scale). In CA, the goals reflect specific community values, for example, ameliorating the effects of inequitable food access; creating job training, educational or therapeutic opportunities for those involved; reclaiming community identity; or exercising active self-determination. Given those types of goals, you most accurately would guess that these sorts of operations are commonly the projects of nongovernmental organizations (NGOs), schools, universities, municipalities, or partnerships and permutations of any of the above. In all cases, I believe that though the food produced in CA has huge cachet—it validates the existence of the effort and provides a compelling and measurable metric for evaluation—the food grown is not culturally as important as the experience of growing food together. The experience of successfully growing food together is not only the key feature of Community Agriculture; it is also a powerful force for changing the culture as a whole. The values brought to life through Community Agriculture work to create a culture where human-scale food makes sense.

VI. COMMUNITY AGRICULTURE AS A CULTURAL FORCE

I work in Community Agriculture, on site, all season long, and have for nearly twenty years. Our little farm exists by the grace of our community and within the biological parameters of this place. We farm anomalous land—it should be houses or commercial development—as do most all Community Agriculture projects. Because we initially flew under the radar, and then because of the appreciation of the services we provide, our community at first tolerated, then celebrated the presence of our farm. Our continued existence depends on our ability to reliably meet the specific needs of our community, according to the specific parameters of this place. Running afoul of our local culture or climate would be a form of hubris, bringing on a predictable and powerful smackdown from the town or the climatic biology we swim in. Our farm is now as endemic and particular to this place as our local festivals and biologically abides by the preclusions of this valley. In this way, the farm stands in public opposition to industrial culture. If you recall from some of the paragraphs above, industrial culture requires (and creates) placelessness, while our farm materially manifests the specific concerns, culture, and biology of this place. Consequently, our farm, as it is, could exist nowhere else but here.

Our farm, like most outposts of Community Agriculture, grows a diverse array of fruits and vegetables in concert with the season; even here in the north, for most of the year, there is good work to be had. We operate from the end of February to the beginning of November. A university, a school district, a city government, and a nonprofit collaborated to create the farm, and, consequently, college students and community volunteers do the work; and schoolchildren are regularly underfoot, making use of the place as an outdoor classroom. Our growing methods run the spectrum from traditional small-scale sustainable agriculture techniques to many-hands organic gardening. We supply a 100-member CSA, custom-grow produce for our food bank, and distribute to a small handful of emergency food shelters. We have bragged about our positive effect on improving access to high-quality food for our most economically vulnerable. We do this first because it is true, but secondly, I believe, because the metric—pounds of food, as well as the food access goal—are both relatively easy for a person to wrap their head around. In this we are quite like all Community Agriculture organizations. However, if food access is actually our primary goal, we have been horribly inefficient, if not utterly wasteful with resources. We could raise money and just buy a few crops in bulk from a major supplier, work with existing distribution networks, and do much better in terms of a dollars-to-pounds ratio; if we did, that ratio would be more favorable than it is now. Or, if we deemed quality, freshness, and carbon footprint important, we could contract with local growers and our regional farmers' cooperative for food and still be money ahead. If pounds of food was our only goal, then organizationally coordinating a handful of institutional partners, relying on untrained amateurs (college students and community members) for labor, and working on such a small scale (ten acres) all seem a ridiculous way to manage an agricultural enterprise. But production is not the main thing. Even though we brag it up, production, pounds of food grown, is not our main goal. How we work, intention, is as important as meeting our production expectations.

On any given summer weekday, college students, teenagers, and a few middle-aged community volunteers work shoulder to shoulder at our farm. The nature of the work, the literal humility of it, its proximity to the ground, erodes the barriers that typically separate us. When squatting, kneeling, and stooping together in dirty work clothes, we are indeed in it together. The labor itself demands enough concentration to hold a person's focus, but not so much to preclude conversation. In fact, sometimes conversation seems the best part of our crops. This scenario played out over time, day in and day out, sets the stage for a specific kind of knowing one another that does not occur in most other contexts. The humility of the work breeds comfort and informality. I regularly see people laughing, talking politics, relationships, religion, all in dramatically candid ways, and even playing word games, while weeding or harvesting. The work is simultaneously conducive to both social comfort and productive effectiveness—together we get the job done and you can see results unfold in real time. Ten people weeding a bed of carrots can do a lot of damage quickly, even if as individuals they work slowly. A group like that, working together over time, as the students and volunteers do at our farm, jells quite quickly. They internalize their effectiveness as a community and also recognize, as individuals, their contributions. This combination of the sense of both individual power and belonging to a community, all while creating real, tangible, and valuable results, is transformative. The process lights people up from within, they walk taller, speak of the farm as "ours," and take tremendous pride in their work—all this from a volunteer experience, a class they are paying to be part of, or in some cases, even a therapeutic juvenile justice activity. The experience of successfully growing food together, not the food itself, is the best thing we produce, and ultimately, I believe, the most powerful.

VII. COMMUNITY AGRICULTURE AS A SOURCE OF SOCIAL CHANGE, AND THE EXCITING CONCLUSION

The argument outlined so far goes like this: the industrial food system grips our world so mightily because it fits hand in glove with our culture. Though we have made fantastic strides creating elements of a human-scale food system, we have not slowed the growth of the mainstream industrial food system, and we cannot, because we have an industrial (production-oriented) culture. Our production-oriented culture powerfully bends the shape of our work lives to meet its needs. Consequently, actions that reflect a high degree of intention—how we get things done, as opposed to the degree to which we are productive—do not fit smoothly into the culture. Industrial agriculture meets the production-oriented needs of our culture, and human-scale agriculture cannot. The drive-through facilitates a nine-hour workday bookended by a commute, whereas whole food, grown, chosen, and prepared with care, does not. The values inherent in human-scale agriculture embody concerns for careful process and specific people, communities, and landscapes—these values are anti-industrial. They struggle to get much of a toehold in an industrial context. Our hard-won, yet limited gains in the effort to undo industrial agriculture reflect the difficult nature of this effort. This work swims against an immense and relentless incoming tide. To extend the metaphor, we do not just need to become better open water swimmers in order to create a more widespread human-scale food system; we need to change the direction of the tide. We must change the culture. Allowing the industrial food system to fade away and fostering the replication of human-scale agriculture across the landscape requires the creation of a culture where human-scale agriculture actually makes mainstream cultural sense. This is the point where Community Agriculture enters the discussion.

We have decided, collectively, to make production our highest value. We have designed our workplaces, farm fields, even the structure of our lives, to maximize production. This is an industrial ethic. The emphasis on production crosses class boundaries within our work and bends nearly all other facets of life around it. Our culture's relentless emphasis on production comes at the cost of other values, some quite universally heartfelt, like justice, fairness, and care for our places. We have grown distant from those values as a society, yet we decry the specific consequences of their absence (fouled rivers, alienated people, and imploded neighborhoods). Like fairytale characters lost in the dark woods, to find our way out of this morass, we should retrace our steps and follow the breadcrumbs back in the opposite direction. I do not mean recreating the past; I mean literally moving in the opposite direction, away from production as an ultimate value and toward intention.

In a typical production system, the people and places involved, like the rest of the tools, inputs, and investments required, all exist to serve production. These elements have limited individual value and can be swapped out relatively easily for something else, someone else, or somewhere else that will serve the same purpose. In order to maximize production, we have had to build the expendability of people and places into the designs of our workplaces and built environments. We can see the consequences of that expendable treatment of land and people everywhere; expendability is, in fact, a commonality linking many a social and environmental loss.

To follow the breadcrumbs out of these dark woods means to go against that grain—to move in the opposite direction and invest specific people and places with importance. That means designing production systems with as much emphasis on intention, how we do things, as the results of the process. In this case—the growing of food—the intention would be focused on the health of people involved and the consequences to the place. What then would a work environment, a project, a shared activity look like when the biological and cultural parameters of a place actually helped define the enterprise? What happens when we view the specifics of a place as foundational, rather than irrelevant to an effort? In a similar vein, what would the rhythm of work look like, feel like, when the effort serves a specific place and individual people, not the other way around? This is not merely rhetorical. If, dear reader, you remember back a few paragraphs ago, when perusing the definition of Community Agriculture (CA), therein lies the answers to those questions.

You may recall that CA embodies the elements described above. CA exists only by the graces of local culture and biology, and the experience of the enterprise, growing food together, serves the people involved. As I hinted at in the description of CA, when done properly, these farms and gardens engender deep personal investment on the part of the participants, and the places themselves shine with verdant beauty. They can become collective art of a sort, outward manifestations of the will of a group, where the dynamic state of the place reflects an intention of care and the specific qualities of the culture of the people involved. When in the midst of CA, the farmers and gardeners come to own the place, not in a real estate sense, but in that, for a while, the place becomes of them. In a similar way, the people who make such a place together become, for a while, of one another. The activity as an entirety is anti-industrial; the specific people involved and the place itself are essential, as opposed to expendable, even as the effort can be fabulously productive. And here is how it can change culture.

Because of the humble nature of the work, we do not have to wait for legislation or gifts from the powers that be to begin Community Agriculture. We may need only limited permission (depending on the scale) or it can be done guerrilla style (see parts of Detroit and Ron Finley's work in Los Angeles). In these efforts we can act on anti-industrial values (justice, fairness, and care for our places) somewhat immediately. Grow food together now. The act of doing it is the creation of culture, as culture is a manifestation of values, as well as a collection of individual actions. We can, in short, begin to make the world we want, right now, even before it is officially here. Though the boundaries of the work may appear to be just the fence line of the farm or garden, the actual effects are decidedly larger and richer than that.

Done well, Community Agriculture profoundly reaches the hearts of the people involved and even inspires those who witness its unfolding from outside the fence. However small these effects may be in area, mighty will they be on the people involved. Ownership, empowerment, community; these sensibilities transform people, make them powerful, and engender a belief in the validity of themselves and their partners in the effort. The effect of Community Agriculture can spread across people's lives into other parts of their existence that may seem separate from their work in the garden. The power of the experience can instill a belief in the possibility of self-determination, in the creation of a personal narrative in accordance with values. A person so infected carries power. The experience, as well as the physical evidence of the success, the productive—yes, productive—beauty of a finely wrought piece of community agriculture, serves as literal inspiration for the legitimacy of anti-industrial values in other work.

When you have felt your ability to make this sort of change in the world, you become a force. And no matter the work you do next, you take this with you. In other words, once having drunk the value-laden Kool-Aid of Community Agriculture, it is awfully hard to become a quiet wheel in the cog of industrial culture. You will believe in your ability to bring better values to light. Now this conclusion will circle back, like all conclusions must. At the beginning I claimed that human-scale agriculture will only make limited gains against its Goliath-like counterpart until we alter the larger culture. Community Agriculture brings forth specific values in the world, and a world where those values—justice, fairness, and care for our places—have found a purpose is actually a world where human-scale agriculture makes sense. Human-scale agriculture not only reflects those values, but also literally physically fits into lives where those values have taken shape.

This whole tirade was about how we need a cultural change in order to see the real growth of human-scale agriculture. Community Agriculture emphasizes intention, puts a focus on the care of specific people and places, and those who come into close contact with this work absorb its effects. As I noted earlier, this absorption of values is not indoctrination, but a version of ownership and an internalized sense of personal power. For a person in this position, where intention has proved more important, soulful, and even effectual than production, exercising those values outside of the garden's fence makes sense. In this way, Community Agriculture pushes the culture and makes way for more widespread support for human-scale agriculture. If we can become accustomed to seeing justice, fairness, and a concern for our places take shape in the world, human-scale agriculture actually makes sense. Obviously, we create culture every day, and there are all kinds of ways to bring values into the world. Community Agriculture projects are not the only way to push for a world where human-scale agriculture makes sense, but it is one way; there are probably many more. As I write, activists all over the globe aikido the system for human-scale agriculture, and they deserve our relentless support. And at the same time, we should not wait for them to make a better world; we can and must make a better world real, right now. Grow food together, in public. Almost anyone can do this; money, higher education, and a powerful social circle are not prerequisites for success. Please note: some of the biggest successes in Community Agriculture are in some of the most marginalized places in the country. Community Agriculture is for everyone. As we create Community Agriculture, we bring forth values into daily life such that a food system built on intention might actually fit into the rhythm of modern life. We can lead lives where human-scale food makes as much practical sense as industrial food does now. Let's get busy on that.

6 Consumer Access and Choice in Sustainable Food Systems

Jane Kolodinsky, University of Vermont

I. INTRODUCTION

Consumers are an essential part of the structure of sustainable food systems (SFSs). Although SFSs have embedded values that are important to multiple cross sections of our society, in order to prosper, SFSs must meet consumer needs and demands by supplying everything from basic food security to high-end value-added foods. SFSs should promote health, environmental sustainability, resiliency, fairness, economic balance, transparency, and diversity of size, scale, geography, culture, and choice.[1] A consumer focus is reflected in a host of indicators used to gauge the success of SFSs, including daily per capita servings of fruits and vegetables consumed, obesity rates, distance to a food retailer, number of farmers' markets, and per capita consumer expenditure on fruits and vegetables. While these indicators are handpicked from much longer lists,[2] they have two things in common that are integral to a thriving SFS: food access and choice.

Our focus is on "final" consumers who make food purchases for their households and create demand in the larger food system. Access and choice decisions also depend on the behavior of other actors in the food value chain. Farmers make purchasing decisions, as do wholesalers and retailers. Governments set policy that ultimately determines what is available for sale. At the same time, we often hear that decisions at the farmer, distributor, wholesale, retail, and policy levels, which influence consumer access and choice, are driven by consumer demand. We acknowledge these complexities in the food system. The focus of this chapter is on where people access food, the choices available at those access points, and how consumer decisions can foster or inhibit sustainable food systems.

II. PLACES PEOPLE ACCESS FOOD

This chapter discusses gardens (home and community), community-supported agriculture (CSA), farmers' markets, community stores (general stores and independent grocery stores), supermarkets and superstores, and institutional purchasing. Understanding these individual components and their potential to facilitate or thwart the growth of SFSs may inform future thinking and action to strengthen SFSs. Potential solutions and suggested future directions are offered in closing.

III. HOME GARDENS

The first supermarket supposedly appeared on the American landscape in 1946. That is not very long ago. Until then, where was all the food? Dear folks, the food was in homes, gardens, local fields, and forests. It was near kitchens, near tables, near bedsides. It was in the pantry, the cellar, the backyard.
—*Joel Salatin*

The venues closest to home where consumers can produce food themselves are individual or, as they are sometimes called, kitchen gardens. These gardens have supported family and household food security since prehistoric times. An important subset of these gardens to note are those that provide the sole support for family food needs, which are considered subsistence gardens.

While gardens across the globe may differ in size, form, and function, home gardens are characterized by their location adjacent to the family home, close association with family activities, and cultivation of diverse crops that meet family food needs on a reliable basis.[3] Because of these characteristics, home gardening in the international arena is viewed not only as a way to improve food security and nutrition by making a variety of fresh produce accessible, but it is also recognized as a way to enhance income and rural employment through additional food production; gain environmental benefits from recycling water and waste nutrients; control shade, dust, and erosion; and maintain or increase local biodiversity.[4] These are, of course, all goals of SFSs.

In the United States, home gardens became less central to household food security during the nineteenth century as the country developed away from an agrarian to an industrialized society. Although home gardens remained an important feature of American culture, they did not regain real prominence until the food shortages of World Wars I and II. In 1943, canned foods were rationed and the secretary of agriculture called for the creation of eighteen million "victory gardens," which were touted as symbols of a strong American work ethic and meant to promote a picture of the "good life" for suburbanites.[5] More than one-third of U.S. produce during this period came from victory gardens, which effectively reduced pressure on the "industrial" food supply.[6] As canned goods were removed from ration lists, the number of victory gardens declined.

Today home gardens are again on the rise. In 2009, approximately 90 percent of American homes reported engaging in some type of gardening.[7] More than a third of gardening households have reported that economic recession motivated them to garden either very much (14 percent) or a fair amount (20 percent).[8] The top reasons for producing food at home through home gardens included access to better-tasting food (58 percent), saving money (54 percent), better food quality (51 percent), and increased assurance of food safety (48 percent).

Local and national attention continues to focus on home gardening as a way to access fresh produce. From Iowa's "Plant. Grow. $ave." social marketing campaign[9] to the Garden Media Group's list of trends[10] that will sell home gardens, home gardening is back as a way to provide consumers with direct access to a variety of food.[11] Bloggers are providing the general public with reasons for home gardening, such as improving health, saving money, reducing environmental impact, getting exercise, enjoying better tasting food, improving food safety, and reducing food waste.[12] Perhaps the most visible recent endorsement for home gardens has come from First Lady Michelle Obama. The White House vegetable garden has attracted media attention and offered the U.S. populace the view that growing a garden at home does not have to be difficult or expensive, can offer a fun way for families to work together and exercise, and can bring healthy food into household meals.[13]

Yet, the question remains: Can home gardening really be a force in supporting sustainable food systems in the United States? Since the early 1990s, academic research in the United States has been amazingly quiet on the benefits of home gardening to Americans and its relationship to SFSs. From a food access perspective, home gardens may only contribute to the availability of fresh produce at the subsistence level and therefore only provide a limited increase in food access.

It is important to note that, because there are no expenditures on the end product (homegrown food) by home gardeners, the value of homegrown produce and value added through food preservation does not count in U.S. gross national product (GNP) other than through the inputs purchased for the garden including things such as gardening supplies, soil amendments, and seeds. This may be one reason for a lack of data and limited current research on the benefits of home gardening for U.S. consumers mentioned above.

Overall, home gardening is limited by geography, climate, skills, space, and time availability. At the same time, increases in home gardening by Americans, even if it means just dabbling, plays a large role in supporting sustainable food systems by increasing awareness of fresh food and consequently increasing demand for fresh food produced by others and sold at other venues. The recognition that fresh food can be hard to grow, but tastes better than some of the alternatives, can drive market demand and contribute to the growth and success of sustainable food systems over the next ten years. There is also the possibility that home gardening spurs other economic activity, such as purchased inputs necessary for food preservation including canning and freezing equipment and supplies.

In the future, researchers interested in home gardening should include a line of questioning in their surveys about whether and how home food gardening impacts demand for food purchased, as there is limited evidence of this in the literature.[14] Lines of questioning might also include investigating whether "talking the talk" of home gardening via social media can help build momentum to keep the values important in SFSs salient in the minds of more and more people.

IV. COMMUNITY GARDENS

Gardens, scholars say, are the first sign of commitment to a community. When people plant corn they are saying, let's stay here. And by their connection to the land, they are connected to one another.
—Anne Raver

Community gardens are pieces of land that are collectively gardened by a group of people who live near each other. Unlike public parks, typically managed by professional staff, community gardens are usually managed by community members, or on occasion local governments or community-based organizations. They can be located on public or private land, which may or may not be split into individual plots. Some community gardens operate according to a variety of rules and regulations, and require members to pay dues.[15]

Community gardens have a long history dating as far back as 100 BCE in the United Kingdom and the small Celtic fields of Lands End, Cornwall, where community gardens are still in use today. During the reign of Elizabeth I (1558–1603), under the feudal system, manorial "common" lands were enclosed and "commoners" were compensated with "allotments" of land attached to tenant cottages. It was not until 1908 that the Allotment Act of Parliament established a legal requirement for local authorities to meet community demand for gardens.[16]

In the United States a program of allotment gardens began in the late 1890s to help meet the needs of families devastated by the effects of economic depression. During World War I, in both the United Kingdom and United States, severe food shortages again triggered the need for community gardens in addition to home gardens. In the United States this involved some 20 million gardeners producing 44 percent of the U.S. supply of fresh vegetables during the war years.[17]

Community gardens provide local consumers with enhanced access to produce and support public health by promoting good nutrition and physical activity. Like home gardens, they enhance individuals' connections to the environment and may contribute to neighborhood revitalization. In addition, community gardens also support strengthened relationships, improve social cohesion, and help build social capital.[18] Recognition of these benefits may have triggered broader consumer interest in community gardening. A 2009 report issued by the National Gardening Association indicated that approximately five million households had expressed interest in participating in a community garden near their home, compared to nearly one million households already involved in community gardening.[19]

Community gardens provide the same type of support for SFSs as home gardens do, but the support is perhaps amplified by the social nature of the endeavor. People create networks while socializing at their shared garden, exchanging perspectives about the food they are growing and (perhaps unknowingly) SFSs in general. These networks are built not only in person but also through social media. On August 1, 2014, a Google search of "community garden blogs" returned 177,000,000 results.

Clearly, people are sharing knowledge about their community garden experiences, but as with home gardens the amount of food grown may only provide produce for those growing it; and only the economic value of gardening inputs, not actual food output, is counted in GNP. In another parallel to home gardens, access to food is limited by geography, climate, skills, and the time commitment necessary to home-produce food.

V. COMMUNITY-SUPPORTED AGRICULTURE (CSA)

We're talking about a farm and a farmer in relationship with the people who are going to eat from that farm.... [CSAs represent] a spiritual renewal of agriculture.
—Allen Balliet, CSA advocate

A CSA is an economically accessible and typically environmentally friendly model of agriculture. CSAs represent commitments between farmers and consumers. Consumers pay to become "members" of a farm in exchange for a basket of agricultural products, made available at a time interval determined by the CSA. Typically, consumers pay up front for membership, providing the farmer or farmers with the capital they need to run a stable agricultural enterprise. Other models of payment have evolved over time, including work in lieu of payment, acceptance of Supplemental Nutrition Assistance Program benefits (SNAP),[20] and installment payment plans.

The system has benefits for both the producer and consumer. A guaranteed revenue stream for farmers reduces their risk as consumer payments are made regardless of the amount and variety of product provided. In bountiful years, consumers often find themselves with an overabundance of food, decreasing the price per unit of food. In the United States, CSAs have typically offered fruits and vegetables, but the type of food available is not restricted to produce. There are CSAs that offer dairy, meat, poultry, fish, eggs, and baked goods. In scarce years, consumers receive smaller amounts of food, or even nothing in the event of a disaster, such as a flood or drought.

Historically, CSAs began operating in the United States in the 1980s, following an earlier European tradition. CSAs originated in Europe (Germany and Switzerland) and Japan during the 1960s.[21] In Japan, the concept of "buying clubs" or "buying groups" began when a lack of food safety plagued the Japanese food system. Housewives in the Tokyo area approached dairy farmers offering to pay up front for milk that would be produced without chemical or synthetic input, thus beginning the "Seikyou"[22] movement. Seikyou then extended to other agricultural products such as vegetables, eggs, meat, and later to consumer goods beyond food and to services. While products out of Seikyou are not always local, they are produced and consumed with a focus on social and environmental concerns. The Japan Organic Agriculture Association (JOAA) was established in 1971, contributing to the foundation of the "Teikei"[23] system, which links consumers to farmers—creating an alternate, direct, food distribution system.[24]

Almost concurrent with the Japanese movement, in 1968, Carl-August Loss and Trauger Groh established Community Land Trusts in Germany. The premise of the German Land Trusts effort was to promote biodynamic farming and the small-scale economic system promoted by Rudolf Steiner. Steiner, an Austrian philosopher and social reformer, promoted ecological and sustainable agriculture (he also founded the Waldorf educational system). Biodynamic farming principles, which promote ecological, social, and economic sustainability, were essentially the precursors of modern organic agriculture. Later, in 1984, after investigating how Steiner's ideas were being implemented, Jan Vander Tuin founded a CSA in Switzerland.[25]

The modern CSA movement in the United States is attributed to the European biodynamic agricultural tradition. In 1986, two CSAs were started independently: Indian Line Farm in Massachusetts and Temple-Wilton Community Farm in New Hampshire. In both cases, these founding farmers had emigrated from Europe where they had previously worked on biodynamic farms. These first CSA farms meshed Steiner's small-scale agriculture and biodynamic farming ideas with E. F. Schumacher's "small is beautiful"[26] economic approach (in direct contrast to "bigger is better" approaches). By the 1990s it was estimated that 60 CSAs were operating across the United States. Today, there are more than 13,000 CSA farms in North America, most of which are in the United States. Regional distinctions characterize CSA farms. In the Northeast farms tend to be smaller, with strong core groups of active consumer members. In the Midwest, CSA farms tend to be driven by producers interested in direct markets. In the West (California), CSA farms are very large. The largest CSA farm in California has 13,000 members.[27] Most recently, food hubs—simply defined as food aggregators, processors, and/or distributors—have started to offer CSA programs.

CSA membership moves consumers away from being producers of food to being partners in food production. Because a monetary transaction occurs, the value of food sold through CSAs is reflected in GDP. Consumers have less of a time commitment compared to gardening, but still must (in most cases) add an additional shopping stop, unless the CSA share is home delivered. As with community gardens, networking and "talking food" can further demand for knowledge about and positive attitudes toward SFSs. Yet, even though there has been an increase in subsidized shares, the ability to work on the farm for a reduced price, and the increasing acceptance of SNAP benefits at some CSA farms, access to food through CSA membership continues to be a challenge for many limited-income people. The reach of CSA farms is limited by geography and the number of members a farm can support.

VI. FARMERS' MARKETS

We've had an explosion of farmers' markets. . . . By shopping at a farmers' market, you support local agriculture, which has a great many benefits. You keep farmers in your community. You keep land from being sprawled with houses and shopping

centers. You have the experience of shopping in the farmers' market, which is the new public square. You support a lot of values when you shop at the farmers' market.
—Michael Pollan

Farmers' markets are common areas where multiple farmers gather regularly to sell a wide range of produce directly to consumers. They can consist of a few stalls or several city blocks and often include value-added products, like samples of food, arts, and crafts. Farmers markets came into existence as a way for agricultural producers to sell surplus crops, and some consider them the flagship of the local food movement.[28] After the passage of the Farmer-to-Consumer Direct Marketing Act of 1976,[29] the number of farmers' markets in the United States increased dramatically.[30]

Farmers' markets allow consumers to buy from multiple local producers at a central location on a regular basis. Incentive programs offered at farmers' markets can increase the purchase and consumption of affordable locally grown produce, benefiting low-income consumers and farmers. For example, in 2010, public nutrition support programs such as WIC[31] and Senior Farmers' Market Nutrition programs served nearly 900,000 seniors and 2.15 million WIC participants,[32] increasing both groups' access to farmers' markets in addition to traditional grocery stores. Some private organizations are encouraging low-income consumers to use their nutrition assistance benefits at farmers' markets by matching benefit amounts, allowing consumers to purchase more produce. Wholesome Wave, a nonprofit organization, offers a "Double Value Coupon Program," which served 40,000 customers in 300 farmers' markets in 25 states in 2012. Fair Food Network in Michigan, Roots of Change in California, and Market Umbrella in New Orleans run similar programs, among others.[33]

As with CSAs, the transactions made at farmers' markets count in our national income accounts. Prices charged tend to be higher than at CSAs, often providing farmers with higher incomes. SNAP benefits and other assistance programs such as "Double Up Food Bucks" are increasing the ability of those with lower income to access produce sold at farmers' markets. But direct to consumer sales do not reach all consumers. McEntee summarizes barriers consumers face with both CSAs and farmers' markets including the time it takes to travel to pick up local produce, the prohibitive up-front costs of the prepay CSA model, the attitudes of the perceived costs of local foods, the perceived elitism of shopping at a farmers' market, or the very real affordability challenges of food.[34]

Despite limited access for low-income consumers, the overall growth of direct to consumer selling has been tremendous. Direct to consumer food sales, defined as "edible farm products for human consumption," grew from $404 million to $1.2 billion between 1992 and 2007. The 2012 Census of Agriculture indicates this growth was twice as large as total agricultural sales.[35] In 2014, beginning farmers with less than ten years of experience represented 22 percent of all direct to consumer sales.

VII. GENERAL AND GROCERY STORES

If you live in a rural community, you understand that our grocery store is arguably one of the most important businesses in town. Our store means more than just ready access to healthy food. Rural grocery stores provide jobs and generate tax revenue. Without a local grocery, the revenue that our food purchases generate goes elsewhere.
—*Center for Rural Affairs*[36]

While direct to consumer food access has grown and is forecast to continue to grow, the majority of food for home consumption is accessed at retail outlets and restaurants. Eight out of the "Top 10" trends for 2014 by the National Restaurant Association are related to SFSs and the measures noted in the introduction to this chapter: locally sourced meats and seafood, locally grown produce, environmental sustainability, healthful kids' meals, hyperlocal sources (e.g., restaurant gardens), children's nutrition, sustainable seafood, and farm/estate–branded items.[37] Demand for local food is also driven by values associated with SFSs, including supporting local economies, providing healthier alternatives, and contributing to sustainability. In addition, consumers perceive more choice.[38] These trends move us from direct to consumer sales venues to more mainstream retail stores and restaurants. The A. T. Kearney study also found that 30 percent of consumers will switch their shopping venue if local food is not available at their preferred grocery store.

Historically, community stores (general stores and local grocery stores) have been important centers for trade and social gathering, especially in rural communities. General stores are retail operations that sell a wide range of consumer goods, including groceries. In the United States, general stores descended from trading posts, where bartering was frequent since money was scarce. Storekeepers sometimes extended long-term credit to consumers, due to the seasonal nature of returns on agricultural produce. As key consumer and community hubs, general stores flourished throughout the nineteenth century. They declined rapidly in the twentieth century, particularly after the 1920s, and were succeeded by specialty stores that handled a relatively narrow product range or particular type of good.

Modern self-service grocery stores evolved from central public markets. Initially located in public streets, markets eventually transitioned to public market houses. As urban areas expanded and transportation options increased, populations moved away from the city center, making central marketplaces no longer adequate. These factors, combined with changes in food processing, led to the introduction of wholesalers.

Wholesalers effectively replaced well-established relationships between sellers and consumers. Corporations became involved in wholesaling, which had a detrimental impact on public markets. Local entrepreneurs capitalized on the opportunity that diminishing public markets created, and began opening small grocery stores, attracting local consumers whose food options were limited by distance and transportation. Generally, small, local distribution worked well for these consumers for some time. Eventually chain stores displaced many locally owned grocery stores.[39]

In rural areas, having a local grocery store helps attract new residents to a town. Similar to a school, a post office, restaurants, and churches, a grocery store makes a community a more attractive place to live. Grocery stores can also be social places where you run into neighbors in the produce aisle, introduce yourself to someone new in town, or catch up on local happenings with the cashier.

Modern life has challenged the viability of grocery stores.[40] Larger centers of commerce are more accessible by car and many people no longer work in the town where they live, choosing to do their shopping elsewhere as part of their work commute. Community stores are also challenged by the limits of scale when ordering from distributors who are used to purchasing larger volumes. Some delivery truck drivers are not able to justify a special trip to reach an out-of-the-way village center. Retail regulations are more limiting than 200 years ago and put compliance strain on small operations.

Residents in many rural areas, no matter their age or income, are affected by the growing phenomenon of rural "food deserts"—the lack of outlets to purchase food. The most recent data available show that 803 counties in the United States are classified as "low access," meaning half or more of the county population lives ten or more miles from a full-service grocery store.[41] USDA has also classified 418 counties as "food deserts"—meaning that all county residents are ten or more miles from a full-service grocery store—and 98 percent of those counties are rural.[42] Residents who live in a "food desert" have less access to a full range of healthy foods, report less healthy eating, and have fewer healthy people.[43] The irony is that many residents in rural food deserts are located near farms but are not able to access local food.[44]

There is some evidence that smaller, independent grocery stores can be a viable venue for consumer access and choice in SFSs. Desai, Kolodinsky, and Roche, in a report for the Vermont Farm to Plate Network, found that there are opportunities for local producers at these types of stores.[45] While different geographies across the United States will likely support different types of food, the study identified opportunities for value-added local food, including specialty bakery items, wine and spirits, produce, and other value-added general grocery items.

VIII. SUPERMARKETS AND SUPERSTORES

We can get chili peppers from Florida all day long, but at the end of the day that is not necessarily the best model for us. . . . I'm going to pay a higher price in Ohio for peppers, but if I don't have to ship them halfway across the country to a store, it's a better deal.
—Darrin Robbins, Senior Manager for Produce, Walmart

Supermarkets are self-service retail markets where produce is sold along with other food and nonfood household items.[46] Supermarkets emerged in the United States in the 1930s as somewhat utilitarian stores, offering food at low prices in industrial spaces. It was not until the 1940s and 1950s that supermarkets became regular channels for food access in the United States. During the 1950s, supermarkets spread throughout parts of Europe, and in the 1960s they began to emerge in developing countries where consumers had developed enough purchasing power and the ability to store food.

Supermarkets remain the number one place Americans purchase their food with more than half (57 percent) of grocery shoppers considering full-service supermarkets as their primary store. Twenty-nine percent report their primary store to be a supercenter (stores that combine general merchandise with groceries, such as a Super Target or Walmart Supercenter), while the remaining 14 percent report their primary stores to be discount stores (e.g., Target or Walmart), club stores (e.g., Costco or BJ's), limited assortment stores (e.g., Aldi or Save-a-Lot), and dollar stores.[47]

Early twenty-first-century studies pointed to supermarkets as being a preferred mechanism for increasing the availability of local produce.[48] Price/value remains a major driver of grocery purchases and seems to be overtaking location, selection, and quality in terms of importance to the consumer. In a similar observation, among grocery store shoppers, the availability of fresh foods was lowest on the list of reasons to choose one store over another (39 percent).[49] Indeed, research has found that people who shop at farmers' markets compared to super-markets, superstores, and convenience stores were more likely to report consuming fruits and vegetables.[50]

According to a USDA study, however, locally grown food is increasingly important to consumers and grocery stores. Storeowners and managers interviewed as part of the study perceived consumers' interest in locally grown food to be based on their preference for high-quality fresh produce and concerns about the local economy, food safety, chemical use, and genetic engineering. Retailers also believed that local foods were valued and purchased for their social and food quality benefits, including support for the local economy and perceived environmental benefits, freshness, taste, and high quality. Retailers and farmers thought that more local food could be sold if larger grocers sourced greater amounts of local farm products. Consumers reported that the benefits of locally sourced food provide a potential competitive advantage over mainstream food.[51]

Supermarkets such as Wegmans, a regional chain, are beginning to respond to consumer demand for local sustainable food. Wegmans now operates a fifty-acre research farm in Canandaigua, New York, specializing in sustainable organic growing practices, but the farm only supplies produce to two of their eighty stores. Media reports indicate Wegmans is thriving economically, while other chains suffer.[52] Walmart advertising has started appealing to consumers by advertising local food,[53] and has reported that while buying locally can yield savings in terms of transportation costs, the main objective of "going local" is to satisfy changing consumer food preferences.[54] Yet, it has been reported that consumers' trust in large retailers and superstores is low when it comes to local foods.[55]

The entrance of supermarkets and superstores into the SFS discussion motivates some SFS advocates and infuriates others. While these venues can reach a large number of consumers with a wide variety of foods, often at an affordable price, the questions of whether the commodification of sustainable food can really be called sustainable rise to the forefront. Difficulties in distribution, delivering enough product, and farmer incomes are just three examples of relevant concerns. At the very least, larger retailers must be authentic in their representation of local, organic, and other foods that represent SFSs. In order to thrive, retailers must be willing to accept new models of distribution. Media reports indicate farmer frustration with superstore buying practices.[56] One study found that retail buyers must establish visibility within each defined region with regard to price and quantities, and make decisions on local assortments.[57] The recommendations suggest that such practices by larger retailers can support movement toward more SFSs by the food retailing industry.

IX. INSTITUTIONAL FOOD PROCUREMENT

Everything is right about Farm to School: healthy fresh food, enhanced economic opportunity for farmers, and education for children about where food comes from. That's a trifecta!
—*Kathleen Merrigan, former Deputy Secretary, U.S. Department of Agriculture*

It is important to recognize that although organizational procurement of food can be a driver of access to SFSs, and although the public can put pressure on institutions to change buying practices, it is not a direct consumer choice. These food procurement strategies, which can involve schools, colleges, hospitals, and prisons, are commonly referred to as farm to institution (FTI), farm to school (FTS), or farm to hospital (FTH) programs.

Organizations participating in FTI programs often combine experiential education around healthy eating with education efforts about community food systems.[58] For instance, the FTS programs "connect schools (K–12) and local farms with the objectives of serving healthy meals in school cafeterias, improving student nutrition, providing agriculture, health and nutrition education opportunities, and supporting local and regional farmers."[59] Growth of FTI programs has been supported by programs such as the Real Food Challenge on college campuses and the Health Care Without Harm's Healthy Food initiative in health care facilities, as well as the National Farm to School Network in K–12 school settings.[60]

Historically, FTS pilot programs started in the mid-1990s in California and Florida. In 2000, the USDA established the National Farm to School Program, which includes research, training, technical assistance, and grants. According to estimates from the USDA Farm to School census for 2011 and 2012, 40,328 schools participate in a FTS program. This represents 44 percent of U.S. schools.[61] The National Farm to School Network (NFSN) was launched in 2007 by more than thirty organizations working around the FTS movement. It works with schools at all levels of involvement in FTS programs. As of 2014, more than 40,000 schools in all states are participating in the network.[62] These FTS programs are used to improve the nutritional quality of school meals, educate children about healthy diets, and create demand for SFSs in the next generation of adults.

Similarly, in the health care sector FTH programs are used to improve the nutritional quality of meals for people who are ill or injured as well as train/retrain people on what a healthy diet is. These programs not only increase demand for SFS foods within the institution, but as with other FTI programs that include dietary education, they also have the potential to increase the demand for SFS foods once patients return home.

Institutional food procurement programs are gaining momentum in higher education and health care settings. The Real Food Challenge works with universities across the country to "shift $1 billion of existing university food budget away from industrial farms and junk food and towards local/community based, fair, ecologically sound and human food sources by 2020." As of 2014, 157 universities had signed on, representing $78 million of food purchases.[63] Health Care without Harm (HCWH) works with hospitals across the country that want to improve the sustainability of their food services. The Healthy Food in Health Care (HFHC) initiative was started in 2005 and provides education, tools, and resources to support health care facilities. The Healthy Food Pledge has been signed by 548 hospitals and seven food service contractors, and they have committed to increase their purchase of local and sustainable food including fair trade and antibiotic free.[64]

Among others, consumer demand for local and healthy food in institutional meals has been cited as a motivator for FTI programs, especially on college campuses where food services feel more pressure to respond to the demand of their clients.[65] Other motivators for institutions to purchase SFSs include supporting the local economy, supporting local farmers, enhancing farm viability, procuring higher quality and healthier food, and obtaining higher sales or participation rates.[66] Barriers included prices, inadequate kitchen equipment, untrained kitchen staff, food safety concerns, inconsistent quality, and seasonality.[67]

X. CONCLUSIONS AND RECOMMENDATIONS

Although there are many variables to consider in the interplay between consumers and SFSs, we can be sure that consumers are an essential part of the structure and health of sustainable food systems. The push for consumers to partake in SFSs has been strong in recent years, and SFSs can be accessed from different venues that offer choice to consumers:

- Home gardens and community gardens may help drive sustainable food systems by triggering an increase in demand for a variety of fresh and accessible food throughout the system. Home gardens and community gardens also appear to be the most demanding of the SFS venues in terms of consumer time and skills. Since the literature is sparse, researchers should gather empirical evidence so we may better understand whether and how home gardens might strengthen SFSs and consumers' access to food.

- CSAs and farmers' markets provide opportunities for consumers to interact directly with the farmers, but there are numerous barriers such as inability to pay the upfront cost, inconvenient pickup times and farmers' markets times, and perception of higher prices. Access to these venues is also disparate at the national level with most CSA programs and farmers' markets located on the West Coast and in the Northeast.

- Grocery stores, supermarkets, and superstores are the venues where the majority of food for home consumption is purchased, and these stores have been increasing their offering of organic and local food due to the strong consumer demand. The negotiation and purchasing power of supermarkets most likely puts pressures on distributors and farmers to keep prices low, meaning that values of the sustainable food systems such as fairness, transparency, and economic balance are not present. Researchers should explore how values of SFSs can be maintained along the supply chain. A concept that has been examined by researchers is the notion of value-based supply chains, but this concept has mostly been examined in the context of institutional food purchasing. Value-based supply chains "preserve the identity of the farmers and ranchers who raised or grew the product being sold, as well as any environmental, social or community values incorporated into its production."[68]

- Institutional food procurement has been receiving a lot of attention from the USDA, but also from state agencies of agriculture and from nonprofit organizations, as the size of institutions and their purchasing powers provide an opportunity to expose consumers to SFSs while providing sizable markets for farmers. Institutions' needs vary by type, and so do motivations and barriers to participating in SFSs. FTS programs have received the most attention from researchers, and attention should be directed toward other institutions to examine the social, environmental, and economic impact of FTI programs.

Of the seven food access points discussed in this chapter, five are still considered "alternative," while only grocery stores and superstores are considered conventional. Although conventional food access points are making strides in the area of SFSs, they do not deliver on many of the values of SFSs noted in the introduction. The other five venues, including home and community gardens, farmers' markets and CSAs, and institution procurement, continue to be seen as "alternative."

A real desire for change is necessary to move the United States toward more SFSs. Consumer demand is one side of the equation. The more consumers are able to and do participate in food systems that deliver on the metrics associated with SFSs, the more of these access points we should have. However, many of these metrics, including environmental sustainability, availability of healthy fresh foods, payment of fair wages to workers, and the equitable distribution of food, are often in conflict with current measures of success, including a growing GDP and expanding profits. A close examination of the production side of the marketplace, which is beyond the scope of this chapter, is necessary to ensure that consumers have both access and the ability to make choices that support SFSs.

NOTES

1. American Planning Association, *Principles of a Healthy, Sustainable Food System* (2010), http://www.planning.org/nationalcenters/health/foodprinciples.htm. In SFSs, a range of food production, transformation, distribution, marketing, consumption, and disposal practices, which differ in size, occur at various scales (local, regional, national, and global). SFSs are geographically diverse, varying in natural resources and climate. They serve as a means to conserve, protect, and regenerate natural resources, landscapes, and biodiversity to ensure that current consumer food and nutrition needs are met are without compromising the ability of the system to meet the needs of future generations. They support diverse cultures, sociodemographics, heritages, customs, and lifestyles. They also support fair and just conditions for communities and provide equitable access to affordable, culturally appropriate, health-promoting food. SFSs offer economic opportunities for a range of diverse stakeholders across geographic regions. Farmers and workers are provided with living wages. Producers and consumers are able to access information necessary to understand how food is produced, transformed, distributed, marketed, consumed, and disposed of. They are empowered to participate actively in decision making throughout the system.

2. G. Feenstra, C. Jaramillo, S. McGrath, and A. N. Grunnell, *Proposed Indicators for Sustainable Food Systems* (2005), http://coloradofarmtoschool.org/wp-content/uploads/downloads/2013/02/Proposed-indicators-for-sustainable-food-systems.pdf.

3. C. Landon Lane, *Livelihoods Grow in Gardens: Diversifying Rural Incomes through Home Gardens,* Agriculture Support Systems Division of Food and Agriculture Organization of the United Nations, Rome (2004), http://www.fao.org/docrep/006/y5112e/y5112e03.htm.

4. D.H. Galhena, R. Freed, and K.M. Maredia, *Home Gardens: A Promising Approach to Enhance Household Food Security and Wellbeing* 2 Agriculture & Food Security (2013), http://link.springer.com/article/10.1186/2048-7010-2-8/fulltext.html.

5. Mann Library, Cornell University, *Harvest of Freedom: The History of Kitchen Gardens in America* (2014), http://exhibits.mannlib.cornell.edu/kitchengardens/index.htm.

6. S. A. Kallen, The War at Home (2000).

7. B. Butterfield, *The Impact of Home and Community Gardening in America* (2009), http://www.garden.org/articles/articles.php?q=show&id=3126.

8. National Gardening Association, *The Impact of Home and Community Gardening in America* (2009), http://www.gardenresearch.com/files/2009-Impact-of-Gardening-in-America-White-Paper.pdf; Garden Media Group, *Garden Media Group Reveals Its 2013–14 Garden Trends Report: Finding "Bliss" by Channeling the Forces of Nature* (2014), http://www.gardenmediagroup.com/clients/client-news/278-garden-media-reveals-its-2013-14-garden-trends-report.

9. Cultivate Iowa, *Plant. Grow. $ave.* http://www.cultivateiowa.org/donate-produce/.

10. Garden Media Group, *supra* note 8.

11. Butterfield, *supra* note 7.

12. L. Barnes and N. Nichols, *The Benefits of Growing Your Own Food* (2010), http://www.sparkpeople.com/resource/nutrition_articles.asp?id=1275&page=2.

13. M. Burros, *Michelle Obama Reveals How Her White House Garden Grows*, N.Y. Times, May 29, 2012, *available at* http://www.nytimes.com/2012/05/29/us/politics/michelle-obama-writes-american-grown.html? smid=pl-share.

14. L. Zepeda and J. Li, *Who Buys Local Food?* 37 J. Food Distrib. Research 1 (2006), *available at* http://ageconsearch.umn.edu/bitstream/7064/2/37030001.pdf.

15. American Community Garden Association, *What Is a Community Garden?* (2007), https://communitygarden.org/.

16. D. J. Humphreys, *The Allotment Movement in England and Wales*, 3 Allotment & Leisure Gardener (1996).

17. S. B. Warner Jr., To Dwell Is to Garden: A History of Boston's Community Gardens (1987).

18. J. Twiss et al., *Community Gardens: Lessons Learned from California Healthy Cities and Communities*, 93 Am. J. Pub. Health 1435 (2003); S. Wakefield, F. Yeudall, C. Taron, J. Reynolds, and A. Skinner, *Growing Urban Health: Community Gardening in Southeast Toronto*, 22 Health Promotion Int'l 92 (2007); P. A. Carney, et al., *Impact of a Community Gardening Project on Vegetable Intake, Food Security and Family Relationships: A Community-Based Participatory Research Study*, 37 J. Cmty. Health 874 (2012).

19. Butterfield, *supra* note 7.

20. According to the U.S. Department of Agriculture, Food and Nutrition Service, SNAP is the largest federal program in the domestic hunger safety net, providing nutrition assistance to low-income households in need.

21. S. Martinez et al., *Local Food Systems: Concepts, Impacts, and Issues*, Economic Research Report No. 97, U.S. Department of Agriculture, Economic Research Service (May 2010), http://www.ers.usda.gov/media/122868/err97_1_.pdf; R.L. Farnsworth et al., *Community Supported Agriculture: Filling a Niche Market*, 27 J. Food Distrib. Research 90 (1996).

22. "Seikyou" is an abbreviated form of "Seikatsu Kyoodo Kumiai," roughly meaning "Living Cooperative Union" in Japanese.

23. In Japanese, "Teikei" means "cooperation," "joint-business," or "link-up."

24. J. M. Kolodinsky and L. L. Pelch, *Factors Influencing the Decision to Join a Community Supported Agriculture (CSA) Farm*, 10 J. Sustain. Agric. 129 (1997).

25. *Id.*

26. See E. F. Schumacher's Small Is Beautiful: A Study of Economics As If People Mattered (1973).

27. Kolodinsky and Pelch, *supra* note 24, at 129–41.

28. A. Brown, *Farmers' Market Research 1940–2000: An Inventory and Review*, 17 Am. J. Alt. Agric., 167 (2002).

29. The intent of the act was to increase domestic access and consumption of locally and regionally produced agricultural projects, and to develop new market opportunities for farmers. The act supported the development, expansion, and provision of outreach, training, technical, and other assistance for farmers' markets, roadside stands, CSAs, and other direct to consumer opportunities.

30. Brown, *supra* note 28, at 167–76.

31. WIC is the Special Supplemental Nutrition Program for Women, Infants and Children, a federal assistance program of the Food and Nutrition Service of the U.S. Department of Agriculture for health care and nutrition of low-income pregnant women, breastfeeding women, and infants and children under the age of five.

32. U.S. Department of Agriculture, *Know Your Farmer, Know Your Food Compass* (2012), http://www.usda.gov/documents/7-Healthyfoodaccess.pdf.

33. Healthy Food Incentives, *Cluster Evaluation, 2011 Final Report* (2011), http://www.fairfoodnetwork.org/sites/default/files/HealthyFoodIncentives_ClusterEvaluation Report_2011_sm.pdf.

34. J. McEntee, *Contemporary and Traditional Localism: A Conceptualization of Rural Local Food*, 15 Local Env't: The Int'l J. Justice & Sustainability (2010).

35. Debra Tropp, *Why Local Food Matters*, National Association of Counties Legislative Conference, March 2, 2014, http://www.ams.usda.gov/AMSv1.0/getfile?dDocName=STE LPRDC5105706; Census of Agriculture, *Beginning Farmer's Report*, ACH12-5/June 2014, http://www.agcensus.usda.gov/Publications/2012/Online_Resources/Highlights /Beginning_Farmers/Highlights_Beginning_Farmers.pdf.

36. Center for Rural Affairs, *Saving the Small Town Grocery Store* (2010), http://www.cfra .org/renewrural/grocery.

37. National Restaurant Association, *What's Hot for 2014* (2014), http://www.restaurant .org/News-Research/Research/What-s-Hot.

38. A. T. Kearney, *Buying into the Local Food Movement*. Ideas and Insights (2013), http:// www.atkearney.com/paper/-/asset_publisher/dVxv4Hz2h8bS/content/id/710104.

39. J. M. Mayo, The American Grocery Store: The Business Evolution of an Architectural Space (1993).

40. J. Bailey, *Rural Grocery Stores: Importance and Challenges* (2010), http://files.cfra.org /pdf/rural-grocery-stores.pdf; P. Clark, L. Tsoodle, and D. Kahl, *Rural Grocery Sustainability Project Owner Survey*, Kansas State University, Center for Engagement and Community Development (2010).

41. L. W. Morton and T. C. Blanchard, *Starved for Access: Life in Rural America's Food Deserts*, 1 Rural Realities 1 (2007).

42. *Id.*

43. J. Bailey, *supra* note 40.

44. Morton and Blanchard, *supra* note 40, at 1–10; J. McEntee and J. Agyeman, *Towards the Development of a GIS Method for Identifying Rural Food Deserts: Geographic Access in Vermont, USA*, 30 Applied Geogr. 16 (2010).

45. Desai Sona, Jane Kolodinsky, and Erin Roche, *Selling Local Food Products at Vermont's Independent Grocers* (2014).

46. Merriam Webster Online, May 16, 2014, s.v. "supermarket," def. 1, http://www .merriam-webster.com/dictionary/.

47. Food Marketing Institute, *Access to Healthier Foods: Opportunities and Challenges for Food Retailers in Underserved Areas* (2011), http://www.fmi.org/docs/health-wellness -research-downloads/access_to_healthier_foods.pdf?sfvrsn=2.

48. *Id.*

49. Food Marketing Institute, *Food Marketing Trends* (2012), http://www.icn-net.com /docs/12086_FMIN_Trends2012_v5.pdf.

50. A. Gustafson, K. Moore, S. Jilcott, J. Christian, and S. Lewis, *Food Venue Choice, Consumer Food Environment, But Not Food Venue Availability within Daily Travel Patterns Are Associated with Dietary Intake among Adults, Lexington, Kentucky, 2011*, 12 Nutrition J. (2013), *available at* http://www.nutritionj.com/content/12/1/17.

51. Martinez, *supra* note 21.

52. J. Voight, *As Americans Rush to Fresh Food, Supermarket Chains Follow*, Healthy Business, October 8, 2012, http://www.cnbc.com/id/49101716.

53. *Walmart's Local Food Ad: NYC Mailer Targets Foodies*, Huffington Post, October 3, 2011, http://www.huffingtonpost.com/2011/08/03/walmart-local-food-ad_n_917639.html.

54. M. Bustillo and D. Kesmodel, *"Local" Grows on Wal-Mart*, Wall Street J., August 1, 2011, http://online.wsj.com/news/articles/SB10001424052702304223804576448491782467316.

55. Kearney, *supra* note 38.

56. A. Fentress Swanson, *Can Small Farms Benefit from Wal-Mart's Push into Local Foods?* National Public Radio, February 2013, http://harvestpublicmedia.org/article/can-small-farms-benefit-wal-mart%E2%80%99s-push-local-foods; A. Fentress Swanson, *Small Farmers Aren't Cashing In with Wal-Mart*, Bay Area Bites, February 2013, http://blogs.kqed.org /bayareabites/2013/02/04/small-farmers-arent-cashing-in-with-wal-mart/.

57. Kearney, *supra* note 38.

58. D. Conner, B. Izumi, T. Liquori, and M. Hamm, *Sustainable School Food Procurement in Large K–12 Districts: Prospects for Value Chain Partnerships*, 41 Agric. & Resource Econ.

Rev. 100 (2012); H. Friedmann, *Scaling Up: Bringing Public Institutions and Food Service Corporations into the Project for a Local, Sustainable Food System in Ontario,* 24 Agric. & Hum. Values 389 (2007).

59. National Farm to School Network, *National Farm to School Network* (2013), http://www .farmtoschool.org/.

60. Health Care Without Harm, *Healthy Food in Health Care* (2012), http://www.healthy foodinhealthcare.org/; National Farm to School Network, *supra* note 59; Real Food Challenge, *Real Food Challenge* (2012), http://www.realfoodchallenge.org/.

61. USDA, *supra* note 32.

62. National Farm to School Network, *supra* note 59.

63. Real Food Challenge, *supra* note 60.

64. Health Care Without Harm, *supra* note 60.

65. B. Horovitz, *More University Students Call for Organic, "Sustainable" Food,* USA Today September 27, 2006; J. Perez and P. Allen, *Farming the College Market: Results of a Consumer Study at UC Santa Cruz,* Center for Agroecology & Sustainable Food Systems (2007).

66. F. Becot, D. Conner, A. Nelson, E. Buckwalter, and D. Erickson, *Institutional Demand for Locally-Grown Food in Vermont: Marketing Implications for Producers and Distributors,* 45 J. Food Distrib. Research 99 (2014); D. Bloom and C. Hinrichs, *Moving Local Food through Conventional Food System Infrastructure: Value Chain Framework Comparisons and Insights,* 26 Renewable Agric. & Food Sys. 13 (2011); B. Izumi, D. Wright, and M. Hamm, *Farm to School Programs: Exploring the Role of Regionally-Based Food Distributors in Alternative Agrifood Networks,* 27 Agric. & Hum. Values, 335 (2010); R. Vogt and L. Kaiser, *Still a Time to Act: A Review of Institutional Marketing of Regionally-Grown Food,* 25 Agric. & Hum. Values 241 (2008).

67. Becot et al., *supra* note 66; C. Dimitri, J. Hanson, and L. Oberholtzer, *Local Food in Maryland Schools: A Real Possibility or a Wishful Dream?* 43 J. Food Distrib. Research 112 (2012); B. Izumi, O. Rostant, M. Moss, and M. Hamm, *Results from the 2004 Michigan Farm-to-School Survey,* 76 J. Sch. Health 169 (2006); Vogt and Kaiser, *supra* note 66.

68. Tracy Lerman, *A Review of Scholarly Literature on Values-Based Supply Chains* (May 2012), http://www.sarep.ucdavis.edu/sfs/VBSCLiteratureReview.Lerman.5.31.12_compressed .pdf.

7 The Workers Who Feed Us
Poverty and Food Insecurity among U.S. Restaurant and Retail Workers
Saru Jayaraman, University of California, Berkeley

M ost Americans experience the food system as consumers—eating out at a local restaurant or shopping at their neighborhood grocery store. Besides being the only two consumer-facing segments of the food chain, the restaurant and food retail industries are also two of the largest and fastest-growing segments of the food system. As Figure 1 demonstrates, they are also the only two segments of the food system to exhibit faster growth than the economy overall.[1]

Unfortunately, both of these sectors provide their employees with largely low or poverty wages and little access to benefits. Corporate actors in these two sectors have successfully lobbied to keep the minimum wage for these workers far below a livable standard and to prevent these workers from accessing benefits such as paid sick days. In a terrible irony, the very workers who sell, cook, and serve our food cannot afford to eat themselves, and often touch our food while sick, not by their own choice. As consumers, our food will never be truly "sustainable" as long as it is sold and served under unsustainable working conditions.

This chapter describes the challenges faced by restaurant and food retail workers and the consequent impacts on these workers, their families, and consumers. Information on the restaurant industry is drawn from a decade of research conducted by the Restaurant Opportunities Centers (ROC) United on the industry nationwide. Data on the food retail sector are focused on California and drawn from the most comprehensive study conducted of food retail workers in that state,

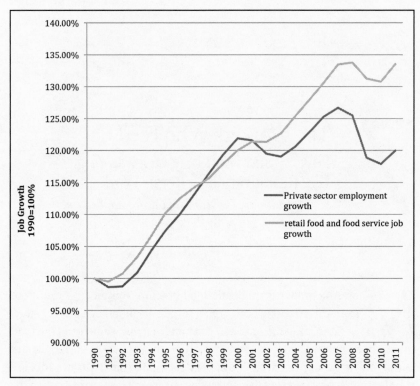

Figure 1. Food Retail and Restaurant Job Growth vs. Private-Sector Employment Growth. *Source: Food Chain Workers Alliance, "The Hands That Feed Us," based on analysis of the Bureau of Labor Statistics, Current Employment Statistics 2012.*

authored by the Food Labor Research Center at the University of California, Berkeley, and commissioned by the United Food and Commercial Workers' Union Western States Council. As the single largest producer of food and the state with the greatest market share of food retail of any state, California presents an important snapshot of conditions faced by food retail workers nationwide.

I. THE RESTAURANT INDUSTRY: NOT LIVING UP TO ITS POTENTIAL

More than 50 percent of Americans eat out at a restaurant at least once per week, and 20 percent eat out two or more times per week,[2] supporting the restaurant industry's continued growth in the midst of the recent economic crises.[3] In fact, the restaurant industry is one of America's two largest private-sector employ-ers, with more than 10 million employees nationwide.[4] It is by far the largest segment of the food chain; more than half of the 20 million workers in the U.S. food system work in restaurants.[5] Census data show that in regions across America, the restaurant industry and service sector clearly presents an increasingly impor-tant aspect of the economy. These jobs are rapidly replacing declining manufac-turing jobs and potentially providing livable wage jobs and career ladders. The National Restaurant Association 2012 industry forecast projected that total indus-try sales would reach a record high of $635 billion, a 3.5 percent increase over 2011, and that one in ten American workers would work in the industry.[6]

As the food system's largest employer, the restaurant industry is not just a bad employer—it is the absolute worst employer in America in terms of wages, benefits, and occupational segregation by race and gender. Unfortunately, despite its growth and potential, the restaurant industry provides largely pov-erty-wage jobs with little access to benefits, pervasive noncompliance with employment laws, and little or no opportunities for career advancement. In 2010, seven of the ten lowest-paid occupations were all restaurant occupations (see Table 1).[7] The median wage for restaurant workers in 2010 was $9.02,[8] meaning that more than half of these workers earned less than the wage of $10.75 that a family of four needs to remain out of poverty.[9]

Indeed, people who earn the minimum wage or less are highly concentrated in the restaurant industry. Thirty-nine percent of all workers making minimum wage or less are in the restaurant industry. Of all workers earning below the minimum wage, almost half (49 percent) are restaurant workers.[10]

The fact that one of the nation's largest and fastest-growing sectors is prolif-erating the nation's lowest-paying jobs can be traced to the power of the National Restaurant Association (NRA), a lobbying group that represents the nation's Fortune 500 restaurant corporations.[11] In 1996, under the leadership of Herman Cain, former candidate for Republican presidential nominee, the NRA struck a deal with Congress saying that it would not oppose a modest increase in the federal minimum wage as long as the minimum wage for tipped workers stayed frozen forever at $2.13 an hour.[12] So the wage for tipped workers such as servers, bussers, and runners has been frozen at $2.13 for the last twenty-two years. Seventy percent of tipped workers are also food servers, who suffer from three times the poverty rate of the rest of the U.S. workforce and use food stamps at double the rate.[13]

Table 1. OES, National Cross-Industry Estimates:
Ten Lowest-Paid Occupations, 2010

Occupational Code	Occupational Title	Hourly Median Wage
35-3021	Combined Food Preparation and Serving Workers, Including Fast Food	8.63
35-2011	Cooks, Fast Food	8.70
39-3011	Gaming Dealers	8.70
35-9021	Dishwashers	8.73
35-9011	Dining Room and Cafeteria Attendants and Bartender Helpers	8.75
39-5093	Shampooers	8.78
35-3031	Waiters and Waitresses	8.81
35-3022	Counter Attendants, Cafeteria, Food Concession, and Coffee Shop	8.83
35-9031	Hosts and Hostesses, Restaurant, Lounge, and Coffee Shop	8.87
39-3091	Amusement and Recreation Attendants	8.87

Source: Tipped over the Edge: Gender Inequity in the Restaurant Industry (Feb. 13, 2012),
http://rocunited.org/wp-content/uploads/2012/02/ROC_GenderInequity_F1-1.pdf.

Women are disproportionately affected by this two-tiered wage system. Seventy percent of tipped workers are women, a majority of whom work at casual restaurants like the Olive Garden, Red Lobster, Denny's, and IHOP.[14] When workers' only hourly guaranteed wage is $2.13, their wages are so low they go entirely to taxes, and workers live completely off their tips.[15] Rita, one of the members of the Restaurant Opportunities Center, told us about the challenges of living on tips alone. "I was a server for 15 years and raised four kids on a server's wages plus tips. Depending on other people to tip you can be the most stressful part of being a server. There were many nights that I didn't even make enough to pay my babysitter. . . . What most people don't realize is that servers don't make the minimum wage like most people."[16]

Like Rita, there are two million mothers working in restaurants; one million are single mothers with children under the age of eighteen.[17] One of the greatest challenges for women living off tips is the vulnerability the situation creates with regard to sexual harassment. Relying on the largesse of customers for their income, female workers' power to stop customers' or coworkers' inappropriate behavior is greatly reduced. As a result, the Equal Employment Opportunity Commission (EEOC) has targeted the restaurant industry as the "single largest" source of sexual harassment claims; 7 percent of American women work in restaurants, but 37 percent of all sexual harassment charges filed by women with the EEOC come from the restaurant industry, which is *more than five times the rate of the general female workforce.*[18]

Ninety percent of restaurant workers report not having paid sick days.[19] Because these workers are told that they will be fired if they take a day off, or they cannot afford to lose a day of work, two-thirds report cooking, preparing, and serving food when sick with illnesses like the flu, H1N1, typhoid fever, and much more. This situation is incredibly dangerous to the public's health; the Centers for Disease Control reports that 50 to 90 percent of outbreaks of norovirus (the winter stomach flu) can be traced back to sick restaurant workers.[20] The CDC also reports that one in eleven restaurant workers reports working with diarrhea and vomiting, which could be cut in half if they had paid sick days.[21] The NRA's unwillingness to spend even a dollar more on workers has created one of the worst public health disasters our nation has faced.

A third major problem these workers face is severe segregation on the basis of race and gender. There are a few livable wage jobs in this industry; waitstaff and bartending positions in fine dining restaurants can earn upwards of $100,000 per year.[22] Unfortunately, workers of color and women face significant barriers in obtaining those livable wage jobs. The Restaurant Opportunities Center (ROC) conducted matched pairs audit testing studies, in which pairs of white workers and people of color were sent into restaurants to apply for fine dining waitstaff positions. We found that white workers were twice as likely as people of color to obtain one of these coveted positions.[23] Because the industry argued that workers of color could not be promoted to waitstaff positions because of their accents, we also sent in pairs of white workers and people of color with accents. We found that for white workers, having a European accent was a bonus, and that for people of color, any kind of accent was a detractor.[24] Not surprisingly, workers of color are concentrated in the industry's lowest-paying positions—bussers instead of waiters and dishwashers instead of cooks—and the lowest-paying segments—fast food instead of fine dining. As a result, workers of color earn a median wage $4 less than that of their white counterparts.[25]

The segregation of women in lower-paid fine dining occupations was borne out in research conducted in New York City, where ROC canvassed 45 Manhat-

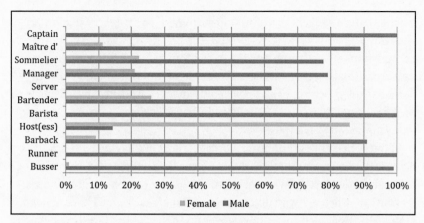

Figure 2. Gender in Front-of-House Restaurant Positions. *Source: Restaurant Opportunities Center canvassing of Manhattan fine dining establishments. Tipped over the Edge, 2012.*

tan fine dining restaurants in 2007. As shown in Figure 2, the results were consistent with our findings that women are underrepresented in the highest-paid positions, such as bartenders, managers, maître d's, sommeliers, and captains. Men held 67 percent of observed highest-paying front-of-the-house positions, while women held only 32 percent. Men held 79 percent of observed front-of-the-house management positions, while women held only 21 percent.[26]

This observation of forty-five Manhattan fine dining restaurants further suggests that the more elite the establishment, the fewer women occupy the highest-paying front-of-the-house positions. American Community Survey data from 2005 to 2009 confirm this observation. During this period, only about 10 percent of front-of-the-house workers in Manhattan restaurants were paid $40,500 or more. However, the front-of-the-house workers earning more than $40,500 per year were more than twice as likely to be male.[27]

Occupational segregation by race resulted in a $3.53 wage gap between white restaurant workers and workers of color in the eight regions, with the median hourly wage of all white workers surveyed in the eight localities being $13.07 and that of workers of color being $9.54.[28] The gap between white and black workers in particular exceeded $4, with black workers earning a median hourly wage of $9. Immigrants and workers of color in the restaurant industry suffer from poverty wages, lack of benefits, and—ironically, as food service workers—lack of access to affordable and healthy food to support their families.

II. SELLING FOOD THEY CANNOT AFFORD: FOOD RETAIL WORKERS IN CALIFORNIA

Unlike the restaurant industry, the food retail segment of the food chain was a historically unionized sector that provided livable wages and working conditions for its workers. Unfortunately, like the restaurant industry, the food retail sector now provides poverty wages and insufficient benefits.

Information in this chapter on food retail workers was drawn from "Hunger and Poverty Amid Aisles of Plenty," the most comprehensive research analysis ever conducted on California's food retail industry. This research shows that while the industry has enjoyed consistent growth and financial health over the past twenty years, the expansion of a low-price, low-cost business model over the same period has had a dramatic effect on workers' wages in this industry. A sector that once enjoyed relatively high unionization rates and wage levels now suffers from high rates of poverty and hunger among workers.

This 2014 study of the California food retail industry was produced by the Food Labor Research Center of the University of California, Berkeley, in collaboration with the Food Chain Workers Alliance and University of California, Davis professor Chris Benner, and was commissioned by the United Food and Commercial Workers Union (UFCW) Western States Council. It was guided by a National Advisory Board comprised of academics and advocates with expertise in the food retail sector and/or the topics covered in this report. The report focuses on data from 925 worker surveys, 20 in-depth interviews with workers, and 20 in-depth interviews with employers conducted in four regions of California: Los Angeles, Southern California outside of Los Angeles, the Bay Area, and the San Joaquin Valley. The data were collected over a nine-month period. This primary research was supplemented with analysis of industry and government data and reviews of existing academic literature.

III. A LARGE AND GROWING INDUSTRY—FEEDING CALIFORNIA

California's food retail industry has shown consistent and robust growth in sales and employment, becoming an important business sector in the state and growing faster than the state economy overall. Between 2000 and 2011, the number of grocery stores in California—the largest segment of food retail establishments in the state—increased by 5 percent, from 9,893 to 10,403.[29] This growth has made the food retail industry an increasingly important source of economic development, impacting the health and well-being of local communi-

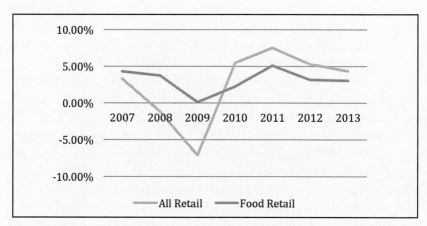

Figure 3. Annual Change in Sales *(U.S. Census Retail Trade Data)*

ties and the state as a whole. California's food retail industry paid workers $7.7 billion in 2011,[30] and generated revenue of $81.4 billion according to the 2012 U.S. Economic Census.[31] Particularly important is the fact that the food retail industry is financially healthy and stable compared to other retail industries. During the 2007 credit crisis and the ensuing recession, the food retail industry maintained continuous sales growth, while the retail industry as a whole experienced extreme volatility, experiencing two straight years of actual declines in sales (Figure 3).[32]

As a result of its relative financial stability, the food retail industry has become appealing to investors. For Wall Street investment banks, food retail's consistent performance means food retailers are good candidates for debt—the stable cash flows allow them to make interest and principal repayments year in and year out, even during recessions, therefore allowing greater capacity for the use of debt in firms' capital structure.[33] Thus, many grocery store chains have become the target of leveraged buyouts.

The food retail industry in California includes three significant segments: traditional grocery stores, specialty food stores, and the grocery share of general merchandise stores such as Walmart, Target, and Costco. With all three segments combined, the study estimates that the food retail industry in California employs 353,390 workers, making up 3 percent of California's total private-sector employment. The largest share of these workers work in traditional grocery stores and specialty food stores, which employ 327,400 workers. With little or no formal credentials required for most jobs in food retail, the industry has been a critical pathway for workers without higher education to achieve a stable

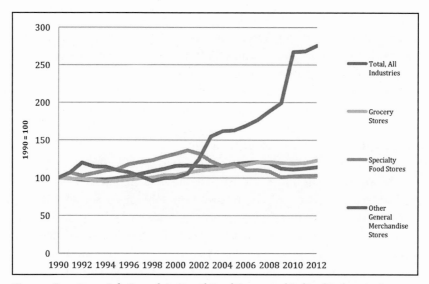

Figure 4. Percentage Job Growth in Retail Food Sector and Related Industries in California, 1990–2012

income. The industry is also an important source of jobs for immigrants and communities of color; in 2011, more than 40 percent of grocery store workers were Latino, and more than 35 percent were immigrants. The industry thus employs more than 150,000 Latino workers statewide.

However, as can be seen in Figure 4, while grocery store jobs have grown faster than overall employment since the year 2000, jobs in general merchandise stores have grown by nearly 200 percent over the same period.

Not all general merchandise stores sell food, but those general merchandise stores that do (namely Walmart, Target, and Costco) have captured a significant share of the grocery market. Walmart currently commands approximately 24 percent of the grocery market nationally,[34] and approximately 7.5 percent in California.[35] It faces significant competition from Target, which spent $500 million in expanding grocery merchandise in 2010. Together, Walmart and Target in particular have captured significant grocery share nationally. Both follow a low-price, low-cost model that reduces quality and specialization in food retail customer service, flattening formerly more developed career ladders in food retail.

In California, Costco has captured a larger grocery share than Walmart and Target and follows a higher-wage, higher-quality business model. Costco pays its workers more than $12 an hour and provides benefits such as paid sick days and access to health care. Nevertheless, despite their smaller share in California, the growth of Walmart's and Target's low-cost model has had an impact on California's food retail sector as a whole. Unfortunately, it appears that the food retail sector in general seems to be following the Walmart model rather than the Costco model. Numerous studies indicate that the growth of this low-cost model creates a downward pressure on wages and working conditions industry-wide.

IV. A DRAMATIC WAGE DECLINE

While employment in California's food retail sector has grown in the past decade, wages have declined. According to the 2010 Census, in 2010 dollars, median hourly wages of grocery store workers—the largest segment of food retail workers—fell from $12.97 in 1999 to $11.33 in 2010, a decline of 12.6 percent. Moreover, the proportion of food retail workers earning poverty wages increased substantially, from 43 percent in 1999 to 54 percent in 2010. This means that, according to the Census, more than half of all food retail workers earn less than the wage needed to reach a low standard of living for a family of three in the Western region if a person works full-time, full-year (2,080 hours).[36]

Meanwhile, overall private-sector median hourly wages increased slightly from $16 to $16.16, an increase of 1 percent. Thus, in 2010, the median hourly wage for grocery store workers was about 70 percent that of the overall workforce. Figure 5 shows the change in average weekly wages for grocery store workers and general merchandise store workers from 1990 to 2012, calculated in 2012 dollars.

As Figure 5 indicates, from 2000 to 2010, grocery store workers suffered a significant weekly wage decline while general merchandise workers experienced a slight weekly wage increase. In sum, grocery store weekly wages fell to nearly the same weekly wage earned by general merchandise store workers. Among grocery store workers, this weekly wage decline was greater for full-time workers and unionized workers. In grocery stores, from 2000 to 2010, part-time workers' wages declined by 2.3 percent, while full-time workers' weekly wages declined by 16.7 percent; similarly, nonunion workers' wages declined by 9.3 percent, while union workers' wages declined by 21.6 percent, more than twice that rate, thus reducing the union wage advantage.

During this same period, historically high unionization rates in food retail declined dramatically. Government data indicate that union grocery store workers earn about three dollars more per hour than nonunion grocery store workers ($13.00 vs. $10.00) (Figure 6) and are slightly more likely to work full-time hours (74 percent vs. 70 percent).[37] Our survey data, reported below, indi-

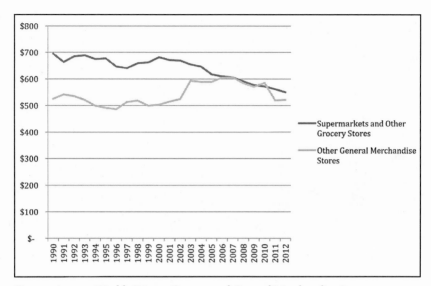

Figure 5. Average Weekly Wages, Grocery and General Merchandise Stores, 1990–2012 ($ 2012)

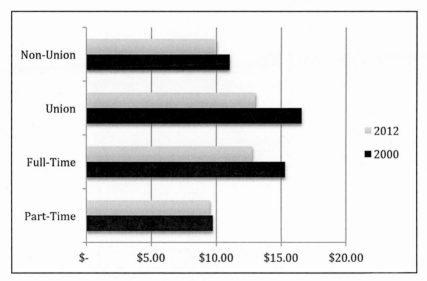

Figure 6. Hourly Wage Change Among Grocery Store Workers by Union and Full-Time Status. *Source: Annual Census (IPUMS).*

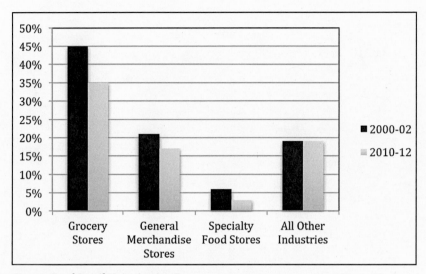

Figure 7. Food Retail Unionization Rates, 2000–2002 vs. 2010–2012. *Source: Current Population Survey—MORG 2000/2002 and 2010/2012.*

cate even greater advantages, with regard to both wages and other elements of job quality, arising from having a union.

However, as described above, market forces such as the growth of the low-cost model in general merchandise stores like Walmart and Target have eroded job quality in the food retail industry as a whole, counteracting the benefits of relatively higher historical unionization rates in this sector.[38] Figure 7 shows that grocery store workers have more than double the unionization rate of general merchandise store workers (35 percent vs. 17 percent), but that both grocery stores and general merchandise stores have each experienced a decline in unionization rates over the last decade.

Figures 4 through 7 combined demonstrate three simultaneous trends: from 2000 to 2010, as the industry as a whole remained financially stable and grew, general merchandise store jobs grew by nearly 200 percent, the unionization rate among grocery store workers declined by almost one quarter (22.2 percent), and wages for grocery store workers dropped 12.6 percent, with full-time and unionized grocery store workers bearing the brunt of the wage decline.

Overall, the food retail worker wage decline from 2000 to 2010 cannot be attributed to an increase in low-wage, part-time work, which stayed relatively constant during that period.[39] Instead, the overall decline in grocery store wages is attributable mainly to two factors: wages for the industry's full-time and unionized workers declined much further than those of part-time and nonunionized workers, and wages of the industry's large number of part-time workers remained much lower than those of full-time workers.

Numerous researchers concur that these simultaneous trends can be explained in large measure by the growth in food retail of the "low-cost" model of nonunion general merchandise stores like Walmart and Target in food retail, which has a negative effect on the union's bargaining power for unionized grocery store workers. As much larger stores require many more workers, general merchandise stores command an increasing share of the grocery labor force. In choosing between emulating the "low-cost" general merchandise model epitomized by Walmart and the high-wage, high-quality general merchandise model epitomized by Costco, unionized grocery store chains appear to be emulating the "low-cost" model, perhaps in part because of pressure they feel from being targeted by Wall Street investors for leveraged buyouts.

As a result, grocery stores may be experiencing pressure to reduce Walmart's and Target's competitive price advantage by mirroring their practices. In fact, researchers Dube, Lester, and Eidlin report that each time a Walmart store opens, it creates incentives for local retailers to offer lower-paying jobs to remain competitive. These researchers estimate that between 1992 and 2000, the opening of a single Walmart store in a county lowered retail wages in that county by between 0.5 and 0.9 percent, and that in the general merchandise sector, wages fell by 1 percent. In addition, they estimate that the greatest burden of this decline in wages falls on grocery store workers.[40]

V. POVERTY WAGES AND FOOD INSECURITY

In surveys, workers described the result of the dramatic wage decline described above. Workers named "low wages" as their greatest issue of concern. Less than 1 percent of workers reported earning a living wage, according to the LLSIL,[41] and almost half (46.4 percent) reported earnings below the poverty line. As a result, food retail workers in California reported double the rate of "low" and "very low" food security as the general U.S. population (Figure 8). In other words, despite the relative financial health of the food retail industry, workers who sell food to eat in California, the largest producer of food in the United States, were twice as likely as the general American populace not to be able to afford sufficient quantities of the food they sell or the healthy kinds of food their families need.

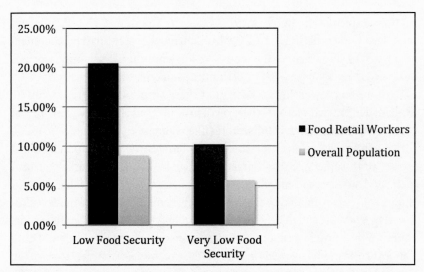

Figure 8. Food Insecurity Among California Food Retail Workers. *Source: Food Labor Research Center Survey Data; USDA.*

VI. THE UNION DIFFERENTIAL

Having a union played a significant role in terms of wages and working conditions for food retail workers.

Table 2. Wages and Working Conditions by Union Status

	Union	**Nonunion**	**All**
Earn Wages Above Regional Poverty Level ($22,458)	69.8%	40.5%	53.6%
Have Health Insurance Through Employer	67.8	35.7	50
Have Health Insurance at All	92.5	68	78
Have Access to Earned Sick Leave	82.4	43.7	61
Received a Promotion	65.7	49.1	57.7
Scheduled Fewer Hours Than They'd Like	30.8	37	34.2
No Lunch Break	12.6	20.6	13.6

Source: Food Labor Research Center, UC Berkeley Survey Data.

As seen in Table 2, unionized workers were far more likely to report earnings above the poverty line and receive promotions than nonunionized workers. Unionized workers also reported having earned sick days at almost double the rate of nonunionized workers. Nonunionized workers were also almost twice as likely as unionized workers to not have a lunch break, even though such breaks are mandated by law.

One of the key areas in which having a union made a difference for food retail workers was access to health care. Unfortunately, some employers' responses to the Affordable Care Act (ACA), which is intended to cover the millions of uninsured workers across America, are creating negative consequences for both union and nonunion workers. Starting in 2015, the ACA assesses a penalty on employers if employees working 30 hours a week are not offered affordable employment-based coverage that meets at least a minimum standard. The employer responsibility provision was created to prevent employers from dropping coverage for their employees who would then receive subsidies through the new marketplaces. Several food retail stores, including Walmart, Target, and Trader Joe's, have already dropped health plans for employees working less than 30 hours a week. The University of California at Berkeley Labor Center estimated that as many as 2.3 million workers nationwide might have their hours cut due to employer reactions to the ACA. The workers most vulnerable to ACA-linked reduced work hours include those working 30–36 hours a week, with incomes below 400 percent of the federal poverty line and a lack of job-based coverage. Retail and restaurant workers account for nearly half of this most vulnerable group.

VII. CONCLUSION

Food retail and restaurant workers live in extreme poverty and are too often unable to afford the food they cook, prepare, serve, and sell. Consumers interested in sustainability and eating ethically must concern themselves with the conditions faced by these millions of workers. Two primary policy recommendations that emerge from the data described in this chapter are to (1) raise the minimum wage for both tipped and nontipped workers; and (2) require employers to provide workers with earned sick leave. ROC and the UFCW are leading campaigns on these issues around the country. Consumers can advocate for these policies both by engaging directly with their legislators and by expressing their values to managers and owners every time they shop for food or eat out. For example, ROC has created a smartphone app—the ROC National Diners' Guide—that consumers can use as a tool to express their values to employers every time they eat out. Changing these industries will result not only in more sustainable working conditions for food retail and restaurant workers, but also more successful employers, and ultimately a better dining experience for all of us as well.

NOTES

1. Food Chain Workers Alliance, *The Hands That Feed Us: Challenges and Opportunities for Workers Along the Food Chain* (June 2012), http://foodchainworkers.org/wp-content /uploads/2012/06/Hands-That-Feed-Us-Report.pdf.

2. Rasmussen Reports (2011) (describing the increase in national food consumption outside the home), *available at* http://www.rasmussenreports.com/public_content/lifestyle /general_lifestyle/july_2013/58_eat_at_a_restaurant_at_least_once_a_week

3. Bureau of Labor Statistics, Current Employment Statistics 1990–2015, Series Ids: CES7072200001 and CES4244500001, Food services and drinking places, and Food and beverage stores. http://data.bls.gov/cgi-bin/srgate.

4. Bureau of Labor Statistics (BLS), *Occupational Employment Statistics, May 2010 National Occupational and Wage Estimates*, http://www.bls.gov/oes/special.requests/oesm10in4 .zip, Food Preparation and Serving Related Occupations.

5. *Id.*

6. National Restaurant Association, 2012 Restaurant Industry Forecast, 2012, at 2–7, 41–6, https://ecommons.cornell.edu/handle/1813/33681.

7. Bureau of Labor Statistics, *May 2010 National Industry-Specific Occupational Employment and Wage Estimates,* Occupational Employment Statistics (May 2010), http://www.bls.gov /oes/special.requests/oesm10in4.zip (The author sorted the data by median hourly wage for all Standard Occupational Classifications).

8. *Id.*

9. Hereafter, unless otherwise stated, "poverty line" or "poverty wage" refers to the income below which a family of four falls into poverty as defined by the 2011 HHS Poverty Guidelines, 76 *Fed. Reg.* 76 3637, 3638 (Jan. 20, 2011). A poverty wage of $10.75 assumes full-time, year-round work.

10. Bureau of Labor Statistics, *Characteristics of Minimum Wage Workers*, Labor Force Statistics from the Current Population Survey Table 4 (2010).

11. Saru Jayaraman, Behind the Kitchen Door (2013).

12. Restaurant Opportunities Centers United (ROC United) et al., *Tipped over the Edge: Gender Inequity in the Restaurant Industry* (Feb. 13, 2012), http://rocunited.org/wp-content /uploads/2012/02/ROC_GenderInequity_F1-1.pdf.

13. *Id.*

14. *Id.*

15. Jayaraman, *supra* note 11.

16. ROC United et al., *supra* note 12.

17. *Id.*

18. *Id.* (emphasis added).

19. ROC United, *Serving While Sick: High Risks & Low Benefits for the Nation's Restaurant Workforce, and Their Impact on the Consumer* (Sept. 30, 2010), http://rocunited.org/wp -content/uploads/2013/04/reports_serving-while-sick_full.pdf.

20. ROC United, *Behind the Kitchen Door: A Multi-Site Study of the Restaurant Industry* (Feb. 14, 2011), http://rocunited.org/wp-content/uploads/2013/04/reports_bkd-multisite.pdf.

21. *Id.*

22. Restaurant Opportunities Centers of New York & the New York City Restaurant Industry Coalition, *The Great Service Divide: Occupational Segregation & Inequality in the New York City Restaurant Industry* (March 31, 2009), http://rocunited.org/wp-content/uploads/2013/04/reports_great-service-divide.pdf.

23. *Id.*

24. *Id.*

25. ROC United, *supra* note 20.

26. ROC United, *supra* note 12, at 19.

27. NWLC calculations of ACS, 2005–2009; Steven Ruggles, J. Trent Alexander, Katie Genadek, Ronald Goeken, Matthew B. Schroeder, and Matthew Sobek, *Integrated Public Use Microdata Series*: Version 5.0 [Machine-readable database] (2010).

28. ROC United, *supra* note 12, at 2–4.

29. United States Census Bureau, *County Business Patterns* (2011), http://www.census.gov/econ/cbp/.

30. *Id.*

31. U.S. Census Bureau, 2012 Economic Census, 445 Food and beverage stores reported revenues of $81.4 billion in California.

32. U.S. Census Retail Trade Data, Time Series Data, Monthly Retail Sales & Seasonal Factors 1992–Present, Retail Total, and Grocery Stores. https://www.census.gov/retail/marts/www/adv44000.txt and https://www.census.gov/retail/marts/www/adv44510.txt.

33. Robert M. Greene, *Industry Analysis: Grocery*, Value Line, http://staging.valueline.com/Stocks/Industries/Industry_Analysis__Grocery.aspx

34. Wal-Mart Stores, Inc. *Form 10-K for fiscal year ended January 31, 2013*, retrieved from SEC EDGAR website http://www.sec.gov/edgar.shtml (The Walmart U.S. segment had net sales of $264.2 billion for 2012 of which 55% were from grocery sales); and U.S. Census Bureau, 2012 Economic Census, 445 Food and beverage stores reported revenues of $620 billion in the United States..

35. Deutsche Bank, *Safeway: Shrinking = Creating Value and Increasing ROIC, Initiating with a Buy* (Sept. 25, 2013).

36. U.S. Department of Labor, Employment & Training Administration, *Lower Living Standard Income Level Guidelines* 78 Fed. Reg. 16,871 (March 19, 2013), *available at* http://www.doleta.gov/llsil/2013/2013llsil.pdf.

37. *Id.*; All comparisons in this section are statistically significant at $p < .05$.

38. Françoise Carré, Chris Tilly, and Lauren D. Appelbaum, *Competitive Strategies and Worker Outcomes in the US Retail Industry*, Institute for Research on Labor and Employment Research & Policy Brief (June 2010).

39. Bureau of Labor Statistics (BLS), Charting the Labor Market: Data from the Current Population Survey (June 3, 2016), http://www.bls.gov/web/empsit/cps_charts.pdf, at 13;

and Bureau of Labor Statistics (BLS), Characteristics of Minimum Wage Workers, 2014 (April 2015), http://www.bls.gov/opub/reports/minimum-wage/archive/characteristics -of-minimum-wage-workers-2014.pdf, at 10.

40. Arindrajit Dube, T. William Lester, and Barry Eidlin, *A Downward Push: The Impact of Walmart Stores on Retail Wages and Benefits*, (Dec. 2007), http://laborcenter.berkeley.edu /retail/walmart_downward_push07.pdf.

41. U.S. Department of Labor *Lower Living Standard Income Level*, *supra* note 36.

Part III: From Federal

Policies to Local Programs

SOLUTIONS FOR A
SUSTAINABLE FOOD SYSTEM

8
A Call for the Law of Food, Farming, and Sustainability

Susan A. Schneider, University of Arkansas

Americal agricultural law and policy have evolved from an early focus on agricultural development and expansion to a current focus on economic and political support for discrete segments of the agricultural sector. This chapter calls for a new approach that can address the unique aspects of agricultural production, while respecting the fragility of the environment and meeting the fundamental need for healthy food. Transforming the special law of agriculture to a new, more inclusive system that focuses on the sustainable production of healthy food is a critical challenge for the future.

Agricultural law, defined as the study of the laws that apply to farmers and the products that they grow, is complex. "Agricultural exceptionalism," that is, the use of legal exceptions to protect the agricultural industry, is pervasive.[1] This term has its American origins in labor law, where agricultural laborers have been historically excluded from many of the protections afforded to other workers.[2] However, the concept is evident throughout U.S. law, with farmers protected from involuntary bankruptcy,[3] exempted from many environmental regulations,[4] and excepted from antitrust restrictions.[5]

Other laws, most notably the federal farm programs, provide unique benefits for farmers, paying billions of dollars to farmers who produce certain favored crops.[6] Additional specialized laws include the federally subsidized system of crop insurance,[7] the special use valuation afforded to farmers for estate planning purposes,[8] the farm loan programs provided to farmers who cannot obtain credit elsewhere,[9] and Chapter 12 of the Bankruptcy Code, a powerful tool available only to "family farmers."[10] Of all industries, only agriculture has its own cabinet department, the U.S. Department of Agriculture (USDA). Over the years, agricultural law scholars have theorized as to how and why this special legal system came about, articulating some of the most persuasive reasons for the existence of a parallel regulatory framework for agriculture. Some defend the special status by tying it to noble societal concerns. Others bemoan the special treatment, linking it to political and economic power.

Are unique agricultural laws a relic from the past? How much of the structure of agricultural law is based upon support for a special interest group, and how much is based on the more overarching needs of a society to feed itself? If support for agriculture is necessary in order to "feed the world," why are so much of our federally supported crops devoted to nonfood uses? The need for food may be an excellent justification for promoting a robust agricultural sector, but in many respects our current agricultural policy has shaped food policy for generations, with our current food system as a result. Is the cart driving the horse?

This chapter argues for the special treatment of agriculture, but not for a status that exempts it from regulation. Rather, it calls for a reconsideration of the framework of agricultural law and the development of support mechanisms that encourage a sustainable food policy. It calls for a policy that supports the economic welfare of the agricultural industry only in the context of the universal societal goal that justifies its special treatment—the production of food. Moreover, it calls for a recognition that "not all food is created equal." Some food serves as healthy fare; other food can actually contribute to health problems.[11] As food production depends on limited natural resources, choices should be made wisely. To the extent that government policies influence the production of food, this influence should be focused on the production of healthy food. "Agricultural law" should be recast as the law of food, farming, and sustainability with the sustainable production and delivery of healthy food to consumers as its central goal.

The need for food is the most rational basis for agricultural law as a unique discipline. Food, as the most basic of human needs, provides a compelling justification for a legal system that nurtures and guides its agricultural sector. A primary role of government is the assurance that its people have sufficient food. Agricultural law scholar Neil Hamilton referred to this as "the fundamental nature of the production of food to human existence" and identified it as one of the primary reasons for the origins of agricultural law as a special discipline.[12]

This food-based agricultural law, however, cannot be driven solely by protectionism or exceptionalism, and it cannot be focused solely on assuring the economic vitality of the agricultural industry. A return to the agrarianism that reconciles the self-interest of farmers with the public good of society should be the hallmark of the new food-based agriculture. Three unique attributes involved in agricultural production are themselves areas of significant public interest: the production and distribution of healthy food, the production of a living product, and the need for the sustainable use of natural resources. These attributes, reflecting the public's interest in agricultural production, should frame the outline of the new food-focused agricultural law.

I. THE PRODUCTION AND DISTRIBUTION OF HEALTHY FOOD

Agricultural production is the primary way that we obtain food—a product that is essential to human health and survival. Both farmers and the public at large have a fundamental interest in the production of healthy foods, in policies that assure the safety of those foods, and in the ready availability of healthy foods to all segments of society. Agricultural law should support and encourage the production and distribution of healthy food.

To date, the production of healthy food has not been the basis for our current agricultural law policies. According to USDA data analyzed by the Environmental Working Group, federal farm policy from 1995 to 2012 directed $256 billion in farm, disaster, and crop insurance subsides to the farming sector.[13] Yet, only a tiny fraction of this support, limited to disaster and crop insurance payments, was provided to those producing crops recognized to be the most healthy—fresh fruits and vegetables.[14]

Field corn has received far more federal support than any other farm commodity produced, over $84 billion in farm program subsidies since 1995.[15] In part due to this support, corn is the most widely produced feed grain in the United States, making up 95.7 percent of U.S. feed grain production.[16] Approximately 80 million acres of U.S. farmland are planted in corn. Most of the corn produced is used for livestock feed, with ethanol as the second largest use. Corn is also processed into many food products including starch, high fructose corn sweeteners, corn oil, and numerous industrial uses.[17]

Cotton production is also extensively subsidized by the federal government.[18] Import restrictions and indirect subsidies support the cane and sugar beet industry, encouraging the production of sugar.[19]

This is not to say that cotton, sugar, and corn should not be produced. Rather, the question is whether federal policy should encourage their production through financial incentives. Energy policies may answer that question in the affirmative for certain nonfood crops, but these policies should not be masked as food policy. Moreover, it is important to recognize the lasting effect of these policies, as subsidies are factored into land values, specialized capital assets are purchased for production, and farming expertise is built around the production of these crops.

While a profitable agricultural industry is essential to assure adequate food production, the interests of the agricultural or food industries should not drive food policy. When federal support is provided, society's interest in the production of healthy food should be the first objective of that support.

Consideration must also be given to the quality of the crops produced. How crops are grown, which varieties are grown, when they are harvested, and how long before they are consumed can dramatically affect the quality of the food produced. Many argue that while industrialized agriculture's focus on quantity, uniformity, and transportability has reduced food prices, it has also resulted in diminished food quality. Anyone with a vegetable garden or who has shopped at a local farmers' market knows the taste difference between homegrown produce and that which is mass-produced, transported long distances, and sold in packages in the supermarket. This anecdotal taste test has received support from scientific testing on fruits and vegetables that shows a reduction in nutrient content over the last fifty years.[20] Another study showed that the focus on high yield as a goal has resulted in diminished nutritional values.[21] Studies that confirm the loss of nutrient value over time postharvest raise questions about our practice of picking produce before it reaches full maturity in preparation for long transit periods and a longer shelf life.[22] These studies all question the generic value that we have placed on our food and the focus of much of our agricultural law policy— produce more for less. They show that there are consequences for this emphasis. As they say, there is no free lunch. Particularly in light of limited natural resources, food quality as well as food quantity must test the efficiency of our production.

The goal of the production of healthy foods must include food safety protections. These protections should not, however, discourage small farming operations and regional food processing centers through regulatory structures that are impossible for smaller operations to meet. Smaller, regional food systems may be key to achieving better food transparency, higher quality products, and better connections between consumers and their food.[23]

Moreover, food safety consideration should not simply focus on pathogens at the point of consumer purchase, but should look at the integrated food production system.[24] The goal of "cheap" food cannot result in food that is unsafe or unhealthy.

The nonpartisan Pew Commission on Industrialized Farm Animal Production provided this type of analysis with respect to concentrated livestock operations.[25] The report described numerous overlapping areas of concern including the high rate of pathogens, the potential for transmission of pathogens from animal to animal and from animal to human, the development of particularly virulent pathogens, and the development of pathogens that are antibiotic resistant.[26] Of particular concern is the industry's dependence on subtherapeutic antibiotics for disease prevention and growth stimulation. Such use can contribute to antibiotic resistance.[27] Serious public health issues are raised.[28]

A new focus on healthy food must give serious consideration to any production method that gives rise to a public health concern. Short-term economic efficiency and the production of low-cost food must be weighed honestly against the long-term externalities including both direct and indirect adverse effects. The government and industry should partner in research that is directed toward the production of healthy food that is produced in a sustainable manner, not simply the cheapest and the fastest production possible.

Beyond the production of healthy food, food should be locally or regionally produced. Our current food system is concentrated and dependent upon the transportation of food products long distances—the transport of feed for livestock, the transport of livestock themselves to feedlots, the transport of crops to processing facilities, and the ultimate "food miles" of products delivered to grocery stores. Nutrition is lost in transit, and crops are selected for transportability rather than nutrition or taste. This system relies heavily on fossil fuel–driven transportation, contributing to climate change.[29] Moreover, as we are now experiencing due to the drought in California, its focus on concentrated production makes our national food security vulnerable to drought and other adverse weather events that affect a specific production region.[30]

The new food-focused agriculture should encourage a diverse and regionally based agriculture that is able to provide local food to customers and retail clients. This can be done through a range of mechanisms, including the direct marketing of products to consumers, the use of local suppliers by retail markets, innovative urban agriculture centers, and the development of regional hubs for distribution. The new food-focused agricultural law should encourage these mechanisms.[31] While the Obama administration, under the leadership of Secretary of Agriculture Vilsack, has begun to address the needs of local agriculture, much more must be done.[32]

II. PRODUCING A LIVING PRODUCT

The agricultural industry is unusual, if not unique, in that it relies on the production of living things. These living things, whether crops or livestock, can grow well or grow poorly. They can die prematurely; they are vulnerable to natural processes and natural forces sometimes regardless of the effort put forth in their production. Farmers are similarly vulnerable in that they are inextricably entwined with the complexities of nature and the fragility of life and death. This gives the industry a special status, and it is a justification for protective treatment.[33]

However, the production of living things also places a heavy responsibility on the agricultural industry. Producing a living product evokes ecological and moral issues that are completely different from the production of an inanimate product. With regard to the production of animals, studies show sentience in livestock species far beyond what we had realized.[34] This gives rise to stronger claims for an ethical duty to care for the animals in a way that is compassionate and not cruel. At present, however, there are no federal animal welfare standards for raising livestock, and state and local standards are met with great resistance from the industry.

Growing plants also gives rise to a responsibility with respect to the impact that the cropping activities may have on other forms of life. New seed forms, such as genetically modified seeds, cross-pollinate with non–genetically modified plants, altering the environment and affecting other farmers' crops.[35] Recent reports confirm that the excessive use of the pesticide glyphosate is causing the development of resistant "superweeds" to form throughout the region of use.[36] We are now discovering that the neonicotinoid pesticides that are applied to seed and later to crops may be a significant part of the problem associated with pollinator death and colony collapse disorder in the honeybee population.[37] The most recent studies indicate harmful effects further up the food chain, with bird species adversely affected.[38]

The way we produce livestock has similar unintended consequences because of the biological processes that are involved. Drugs such as antibiotics, hormones, and growth promotants are given to livestock and are excreted in manure. If runoff occurs, there is a risk of the drugs contaminating waterways. If the manure is used as fertilizer, the drugs may later be found in our foods.[39]

Moreover, to the surprise of many consumers, there currently exists no forum for a consideration of ethical issues regarding the production of food. This was acknowledged when the use of cloned animal meat came before the FDA for approval. The FDA stated that it had no authority to consider ethical issues, but was limited to its authority to consider only the safety of the immediate product sold to consumers.[40]

These are issues that confront not only farmers, but society as a whole. While technological advances should not be discouraged, their implementation must be accomplished in a way that will not adversely impact the health, safety, or integrity of the food system or the environment we depend on, today or in the future. Prior to the implementation of new technologies in agriculture production, the long-term, integrated impact and any unintended consequences must be analyzed. We must recognize that the technologies we apply to living products in the natural environment will have effects beyond their immediate application. And we must provide for coordinated regulation and a more holistic approach to the long-term implications of our decisions.

III. THE IMPORTANCE OF SUSTAINABILITY

Agricultural production is a highly consumptive activity.[41] The agricultural sector uses more natural resources, including land[42] and water,[43] than any other single industry. It is recognized as a major polluter of water[44] and a significant source of global warming.[45] Yet, because it is a dispersed industry with environmental effects that are often only noticed over time and with accumulated impact, it is difficult to regulate.

As land and other natural resources are finite, societal interests in preservation are paramount. In addition to preservation concerns, there is continual competition between potential uses, both within and outside of agriculture. Not only must agricultural law policies assure that a sufficient amount of these resources are devoted to the production of food, this production must be environmentally sustainable, that is, "the cultivation and harvesting of crops must leave the land able to support comparable or greater, future yields."[46]

The true agrarianism described by Wendell Berry should be the goal of agricultural environmental policy:

> Agrarian farmers see, accept, and live within their limits. They understand and agree to the proposition that there is "this much and no more." Everything that happens on an agrarian farm is determined or conditioned by the understanding that there is only so much land, so much water in the cistern, so much hay in the barn, so much corn in the crib, so much firewood in the barn, so much food in the cellar or freezer, so much strength in the back and arms—and no more.[47]

While this is the goal, it cannot be realized by trusting in the stewardship of farmers.[48] Self-interest, short-term goals, and financial stresses provide too much temptation. Agriculture needs a distinct legal scheme that regulates the environmental problems of agricultural production and that rewards sustainable production.

In this regard, environmental policies must be based on the competing interests of agricultural production and environmental protection. Environmental externalities must be recognized so that the cost of production can be accurately determined. Only when long-term environmental costs to society are recognized will there be adequate incentive for the problems to be efficiently addressed. Moreover, when these costs are accurately computed, the profitability of more sustainable farming operations will be recognized. Through a combination of direct regulation, incentives for sustainable practices, and additional research and support for sustainable agriculture, the appropriate balance can be achieved. Incorporated into these policies should be focused farmland preservation mechanisms that protect farmland, particularly prime land in and around urban areas.

Finally a sustainable system of agricultural production requires a consideration of social sustainability. The use of human resources in agriculture, the darkest side of agricultural exceptionalism, must be addressed. A complete review of the agricultural labor laws should be undertaken to reconcile the treatment of farmworkers with the ideal of "our professed belief that honest labor should be justly rewarded."[49]

IV. CONCLUSION

As sustainable agriculture advocate and organic farmer Fred Kirschenmann wrote, "[h]uman health cannot be maintained apart from eating healthy nutritious food, which requires healthy soil, clean water, and healthy plants and animals. It is all connected."[50]

America's history includes a rich tradition of agricultural productivity, and we have all benefited from it. Agricultural laws and policies have supported that productivity through an agricultural exceptionalism based on a recognition of the special attributes of agricultural production. Along the way, however, public interest has taken a back seat to special interest. Farm policy has driven food policy, and farmers have been encouraged to farm in ways that are not sustainable, producing crops that are not good for consumers.

It is past time for agricultural law policy makers to reconsider the direction of agricultural policy and to develop a food-focused agricultural law that is based on the sustainable production of healthy food. It is time for a law of food, farming, and sustainability.

NOTES

1. While this chapter discusses agricultural exceptionalism in the context of U.S. law, the first use of this term is often credited to international trade scholarship, where special exceptions are evident in other countries. Grace Skogstad, *Ideas, Paradigms and Institutions: Agricultural Exceptionalism in the European Union and the United States*, 11 Governance 463, 468 (1998) (explaining that agricultural exceptionalism is based both on the specific interests and needs of farmers and upon the broader national interest in a secure food supply).

2. The most notable current exceptions are that "agricultural laborers" are excluded from the definition of "employee" for purposes of protection under the federal National Labor Relations Act, 29 U.S.C. § 152(3) (2012); and "any employee employed in agriculture" is exempt from the overtime pay requirements of the Fair Labor Standards Act, 29 U.S.C. § 213(b)(12) (2012). A limited exclusion for minimum wage protection still exists under the Fair Labor Standards Act; previously agricultural workers were completely excepted. 29 U.S.C. § 213(a)(6) (2012). *See* Marc Linder, *Farm Workers and the Fair Labor Standards Act: Racial Discrimination in the New Deal*, 65 Tex. L. Rev. 1335 (1987).

3. 11 U.S.C. § 303(a) (2012).

4. See J.B. Ruhl, *Farms, Their Environmental Harms, and Environmental Law*, 27 Ecology L.Q. 263, 293–327 (2000) (describing the "active and passive safe harbors farms enjoy" under environmental law).

5. 7 U.S.C. § 291 (2012).

6. 7 U.S.C. ch. 113—*Agricultural Commodity Support Programs*, §§ 8701–8793; *See* Environmental Working Group, *Farm Subsidy Database Update*, http://farm.ewg.org/. For a review of the history of the federal farm programs, *see* Allen H. Olson, *Federal Farm Programs—Past, Present and Future—Will We Learn from Our Mistakes?*, 6 Great Plains Nat. Resources J. 1 (2001).

7. 7 U.S.C. ch. 36, subch. I, *The Federal Crop Insurance Act*, §§ 1501–1524 (2012).

8. 26 U.S.C. § 2032A (2012).

9. 7 U.S.C. ch. 50, *Agricultural Credit*, §§ 1921–1949 (2012).

10. 11 U.S.C. § 109(f) (2012).

11. *See, e.g.*, Michael Moss, Salt Sugar Fat: How the Food Giants Hooked Us (2013).

12. Neil D. Hamilton, *The Study of Agricultural Law in the United States: Education, Organization and Practice*, 43 Ark. L. Rev. 503, 504 (1990).

13. Environmental Working Group, *Farm Subsidy Database Update*, http://farm.ewg.org/.

14. Jean M. Rawson, *Fruits, Vegetables, and Other Specialty Crops: A Primer on Government Programs*, CRS Report for Congress, RL32746 (Jan. 26, 2007) (noting that "specialty crops

are ineligible for the federal commodity price and income support programs" but explaining other types of USDA assistance may be available including "crop insurance, disaster assistance, and, under certain conditions, ad hoc market loss assistance payments").

15. Environmental Working Group, *Farm Subsidy Database, United States Summary Information*, EWG Farm Subsidies, http://farm.ewg.org/region.php?fips=00000&statename=theUnitedStates.

16. USDA Economic Research Service, *Overview, Corn*, http://www.ers.usda.gov/topics/crops/corn.aspx#.U8FmzI1dV8s.

17. Allen Baker and Heather Lutman, *Feed Year in Review (Domestic): Record Demand Drives U.S. Feed Grain Prices Higher in 2007/08*, USDA, Economic Research Service, FDS-2008-01 (2008), http://usda.mannlib.cornell.edu/usda/ers/FDS-yearbook/2000s/2008/FDS-yearbook-05-23-2008_Special_Report.pdf; see also USDA, Economic Research Service Briefing Room—*Corn*, http://www.ers.usda.gov/topics/crops/corn.aspx#.U8FmzI1dV8s.

18. Environmental Working Group, *Cotton Subsidies*, http://farm.ewg.org/progdetail.php?fips=00000&progcode=cotton.

19. *See* Remy Jurenas, *Sugar Program: The Basics*, Congressional Research Service Report, R42535 (Apr. 1, 2014).

20. *See* Michael Pollan, In Defense of Food: An Eater's Manifesto 118–19, 215–17 (2008); *see also* Donald R. Davis, *Declining Fruit and Vegetable Nutrient Composition: What Is the Evidence?* 44 Hort. Sci. 15 (Feb. 2009) (demonstrating decline in the concentration of certain nutrients in vegetables over the last 50–100 years); Donald R. Davis, Melvin D. Epp, and Hugh D. Riordan, *Changes in USDA Food Composition Data for 43 Garden Crops, 1950 to 1999*, 23 J. Amer. C. of Nutrition 669 (2004).

21. Brian Halweil, The Organic Center Critical Issues Report, Still No Free Lunch: Nutrient Levels in U.S. Food Supply Eroded by Pursuit of High Yields (Sept. 2007).

22. Joy C. Rickman, Diane M. Barrett, and Christine M. Bruhn, *Review: Nutritional Comparison of Fresh, Frozen, and Canned Fruits and Vegetables, Part 1, Vitamins C and B and Phenolic Compounds*, 87 J. Sci. of Food & Agric. 930 (2007).

23. *See* Neil D. Hamilton, *Essay—Food Democracy and the Future of American Values*, 9 Drake J. Agric. L. (2004).

24. *See* Susan A. Schneider, *Examining Food Safety from a Food Systems Perspective: The Need for a Holistic Approach*, 2014 Wisc. L. Rev. 397 (2014).

25. Pew Commission on Industrialized Farm Animal Production, *Putting Meat on the Table: Industrial Farm Animal Production in America* (2009) (a project of the Pew Charitable Trusts and Johns Hopkins Bloomberg School of Public Health, http://www.ncifap.org/). See also E. K. Silbergeld et al., *Industrial Food Animal Production: Food Safety, Socioeconomic, and Environmental Health Concerns*, 19 Epidemiol. S15 (Nov. 2008) (raising many of the same concerns globally); Union of Concerned Scientists, *CAFOs Uncovered: The Untold Costs of Confined Animal Feeding Operations* (2008), http://www.ucsusa.org/sites/default/files/legacy/assets/documents/food_and_agriculture/cafos-uncovered.pdf.

26. Pew Commission on Industrialized Farm Animal Production, *supra* note 25, at 13–16.

27. *Id.*; *See also* Jenny Li and David Wallinga, Institute for Agriculture and Trade Policy, *No Time to Lose: 147 Studies Supporting Public Health Action to Reduce Antibiotic Overuse in*

Food Animals Healthy Food Action, http://www.iatp.org/files/2012_11_08_Antibiotics Biliography_DW_JL_long_hyperlinks.pdf (providing an extensive bibliography of recent studies linking the use of antibiotics in animal production to the development of antibiotic-resistant bacteria).

28. Centers for Disease Control, *Antibiotic Resistance Threats in the United States* 2013, http://www.cdc.gov/drugresistance/threat-report-2013/index.html (stating that [a]ntimicrobial resistance is "one of our most serious health threats").

29. *See generally* Rich Pirog, Timothy Van Pelt, and Kamyar Enshayan, *Food, Fuel, and Freeways: An Iowa Perspective on How Far Food Travels, Fuel Usage, and Greenhouse Gas Emissions*, Leopold Center for Sustainable Agriculture (June 2001), http://www.leopold.iastate.edu/pubs/staff/ppp/food_mil.pdf. *See also* Marne Coit, *Jumping on the Next Bandwagon: An Overview of the Policy and Legal Aspects of the Local Food Movement*, 4 J. Food L. & Pol'y 45 (2008) (discussing popular dissatisfaction with the distance between food production and consumption and the rise of the local food movement).

30. USDA, Economic Research Service, *California Drought* 2014: *Farm and Food Impacts*, http://ers.usda.gov/topics/in-the-news/california-drought-2014-farm-and-food-impacts.aspx#.U8F4MI1dV8s (noting that "because California is a major producer in the fruit, vegetable, tree nut, and dairy sectors, the drought has potential implications for U.S. supplies and prices of affected products in 2014 and beyond").

31. *See, e.g.*, University of Arkansas Community Design Center, *Fayetteville* 2030: *Food City Scenario*, 1 (2013) http://uacdc.uark.edu/work/fayetteville-2030-food-city-scenario-plan?rq=fayetteville%202030%20food%20 (an award-winning design for incorporating local food systems into a city's projected growth, with the long-term goal of providing a significant percentage of the food needs of the city through local production systems).

32. *See* Sarah Morath, *The Farmer in Chief: Obama's Local Food Legacy*, 93 Or. L. Rev. 91 (2014).

33. *See* Susan A. Schneider, *What Is Agricultural Law*, 26 Agric. L. Update 1 (Jan. 2009).

34. *See, e.g.*, the public signing and proclamation by noted scientists, *The Cambridge Declaration on Consciousness in Non-Human Animals*, presented at the Francis Crick Memorial Conference, Cambridge, UK, July 7, 2012 (calling for the public acknowledgment of the conclusive research that establishes that humans are not alone in possessing "the neurological faculties that constitute consciousness as it is presently understood").

35. *See, e.g.*, Eva Hershaw, *Seeds of Discontent: A Texas Organic Cotton Farmer Takes On Monsanto*, Tex. Observer Feb. 7, 2013 (reporting on the contamination of organic cotton by genetically modified cotton seed drift).

36. *See* Jason Koebler, *Herbicide-Resistant "Super Weeds" Increasingly Plaguing Farmers*, U.S. News & World Rep't Oct. 19, 2012, *available at* http://www.usnews.com/news/articles/2012/10/19/herbicide-resistant-super-weeds-increasingly-plaguing-farmers.

37. Penelope R. Whitehorn, Stephanie O'Connor, Felix L. Wackers, and Dave Goulson, *Pesticide Reduces Bumble Bee Colony Growth and Queen Production*, Sci. 351 (Apr. 20, 2012).

38. Caspar A. Hallmann et al., *Declines in Insectivorous Birds Are Associated with High Neonicotinoid Concentrations*, Nature 341 (July 9, 2014), http://www.nature.com/nature/journal/vaop/ncurrent/full/nature13531.html.

39. *See* Susan A. Schneider, *supra* note 24, at 405–16.

40. *Animal Cloning Risk Assessment; Risk Management Plan; Guidance for Industry; Availability*, 73 Fed. Reg. 2923 (Jan. 8, 2008).

41. *See generally* David E. Adelman and John H. Barton, *Environmental Regulation for Agriculture: Towards a Framework to Promote Sustainable Intensive Agriculture*, 21 Stan. Envtl. L.J. 3 (2002).

42. *See* Ruhl, *supra* note 4, at 265.

43. *See, e.g.*, Piet Klop and Jeff Rodgers, *Watering Scarcity: Private Investment Opportunities in Agricultural Water Use Efficiency*, World Resources Institute (Nov. 2008) (reporting that "[a]griculture is by far the biggest water user, accounting for more than 70% of global withdrawals").

44. USDA, Economic Research Service, Agricultural Resources and Environmental Indicators (2006 ed.), Chapter 2.2: Water Quality—Impacts of Agriculture (recognizing agriculture as "the leading source of impairment in the Nation's rivers and lakes, and a major source of impairment in estuaries").

45. Keith Paustian, John M. Antle, John Sheehan, and Eldor A. Paul, Agriculture's Role in Greenhouse Gas Mitigation (2006) (Prepared for the Pew Center for Global Climate Change, reporting that "[g]lobally about one-third of the total human-induced warming effect due to GHGs comes from agriculture and land-use change. U.S. agricultural emissions account for approximately eight percent of total U.S. GHG emissions when weighted by their relative contribution to global warming"). These figures do not include food transportation costs.

46. David E. Adleman and John H. Barton, *Environmental Regulation for Agriculture: Towards a Framework to Promote Sustainable Intensive Agriculture*, 21 Stan. Envtl. L.J. 3, 31 (Jan. 2002).

47. Wendell Berry, "The Agrarian Standard," *in* The Essential Agrarian Reader, the Future of Culture, Community, and the Land, 29–30 (Norman Wirbzba, ed., 2003).

48. J.B. Ruhl, *Farmland Stewardship: Can Ecosystems Stand Any More of It?*, 9 Wash. U. J.L. & Pol'y 1 (2002) (challenging the notion that society can rely on the farmers' inherent stewardship for environmental protection).

49. *See generally* Daniel Rothenberg, With These Hands: The Hidden World of Migrant Farmworkers Today (1998) (providing a compelling description of the underclass of migrant farmworkers and the hand labor they perform as an integral part of our food system).

50. Fred Kirschenmann, *Farming, Food, and Health*, Gleanings, 1 (Summer 2006), http://www.leopold.iastate.edu/sites/default/files/pubs-and-papers/2006-08-food-farming-and-health.pdf.

9 Informational and Structural Changes for a Sustainable Food System

Jason J. Czarnezki, Elisabeth Haub School of Law at Pace University

I. INTRODUCTION

The relationships between food systems, law, and the environment are strong.[1] The ecological costs of modern industrial and large-scale food production are driven by greenhouse gas emissions, fertilizers and pesticides, and food miles,[2] as well as agricultural law. Food choices contribute to the climate crisis, cause species loss, impair water and air quality, and accelerate land use degradation.[3] For example, "An estimated 25 percent of the emissions produced by people in industrialized nations can be traced to the food they eat."[4]

The ecological costs of the modern industrial, carbon-heavy food system are well chronicled. Chemical inputs in the form of fertilizers and pesticides have the potential, through runoff, to pollute groundwater and streams, cause algae blooms and oxygen depletion in waterways, contribute to soil acidification, kill beneficial insects, and potentially poison wildlife and their reproductive systems. Industrial farming techniques such as overtilling, a lack of crop rotation, use of inorganic fertilizers and pesticides, and the agricultural practice of monoculture mine the soil of its natural nutrients, destroy soil biota and its habitat, and increase erosion. And contributing to the climate crisis, fossil fuels remain the single most important ingredient in the modern food system, not only used as fuel for transportation and production of food, but also to produce fertilizers and pesticides.

In an effort to change food choices and inform consumers of the environ-mental impacts of food, I have already argued for creation of an eco-label for food based on an environmental life-cycle analysis from production, to use, to distri-bution, building on existing organic and carbon labeling programs.[5] But improved eco-labeling is only a start, since it only provides information to con-sumers on available food products that are often industrially produced and pro-cessed. It does not directly improve and increase the supply of and access to ecologically friendly food products (though this may do so indirectly due to consumer demand). Both informational regulation that helps influence con-sumer choice and structural changes that provide consumers with better access to better choices are necessary for a sustainable food system to develop.[6]

Thus, in addition to improving labeling schemes to support environmen-tally friendly food consumption, the market of available food products must be improved. Public law and policy drives American food choices and, in turn, fosters environmental degradation. Michael Pollan, author of *The Omnivore's Dilemma*, wrote in an open letter to the next president of the United States during the 2008 campaign season,

> It must be recognized that the current food system—characterized by mono-cultures of corn and soy in the field and cheap calories of fat, sugar and feedlot meat on the table—is not simply the product of the free market. Rather it is the product of a specific set of government policies that sponsored a shift from solar (and human) energy on the farm to fossil-fuel energy.[7]

Legal policies might better support a low-input, more local, and less pro-cessed market. Already significant efforts are underway to build a more com-munity-driven food system that would reduce food miles, decrease consumption of processed foods that contribute to greenhouse gas emissions, and lessen the impacts of chemicals on the environment. While overarching changes in national agricultural law and policy are necessary, beginning with the Farm Bill,[8] second-best solutions like eco-labels and creating new food markets are useful steps. Such steps are even more important given that the organic market is becoming dominated by actors of industrial agriculture, and "the organic sector is coming increasingly to resemble other sectors of commodity-driven agriculture."[9]

This chapter considers legal, theoretical, and practical steps to a more sus-tainable food model. Part I discusses the underlying reasons for problems in the current food system, including those manifested in law, and the perceived ben-efits of creating a new agricultural paradigm. Part II discusses the major agricultural and food programs that have become more common in shaping a different food system model, specifically focusing on direct marketing (for example, farmers' markets and community-supported agriculture) and the organic movement as it relates to small farmers. Part III argues that in order to change modern American food consumption, two changes must take place—increased awareness and increased availability. This chapter reiterates the need to increase the amount of information available to consumers and the consequences of food choices. It further argues that structural changes in the food system are necessary to increase access to sustainable foods by building on current efforts to increase direct marketing by farmers and the number of farmers that are certified, creating better food system planning through state food policy councils and municipal planners, building on existing interests in intrastate and regional efforts supporting local food and local economies, and improving management of existing alternative agricultural distribution and production systems.

II. LEGAL IMPEDIMENTS AND THE NEW AGRICULTURE

The transition away from the modern industrial food system to a different agricultural model has many names—civic, alternative, and new. Professor Thomas Lyson promotes a "civic agriculture," a term that "embodies a commitment to developing and strengthening an economically, environmentally, and socially sustainable system of agriculture and food production that relies on local resources and serves local markets and consumers."[10] An "alternative" food system would incorporate organic foods, eco-labeled foods, direct marketing, fair trade, local foods, farmers' markets, and buying clubs.[11] And a "new agriculture" model could create opportunities to keep farm families on the land and create new farms; promote sustainable farming practices to protect the environment and support profitable farms and communities; build diverse efficient local food systems designed to address local food needs; and create opportunities for people at all levels of the food economy.[12]

The existing industrial food model, heavy on chemicals, fossil fuels, and industrial processing, has been created, in part, by laws that have also impeded the creation of new agriculture. Ironically, these laws have also provided justification for creation of new agricultural and food models.[13] Many legal, policy, and social constraints must be overcome to create a new agricultural model: legitimate concerns about food safety and public health, which often result in regulatory impediments that can overwhelm small farms and processors;[14] the federal government's heavy subsidization of commodity grains through the

Farm Bill; the rising cost of food creating a comparative advantage for industrially produced and commodity-driven foods; the emergence of large agribusiness biotechnology, genetically modified crops, and concentration animal feed organization; and the continued reliance on fossils fuels for food production and distribution.[15]

"New agriculture" attempts to overcome these obstacles in the absence of a fundamental shift in national food and agriculture policy. The new agricultural movement supports a sustainable food system: locally and/or efficiently produced, processed, and distributed foods; an economically viable market for farmers and consumers; and ecologically sound and/or organic production, processing and distribution.[16] New agriculture supports a sustainable food system: increasing direct farm marketing and local food buying, and creating opportunities for new markets and foods; changing the model of institutional purchasing so state and local government can create demand for sustainable foods; and supporting eco-labeling and food education programs so consumers can act on their concerns to influence changes in food and farming practices.[17]

Changing what we eat and the way we eat will require significant and intentional modifications in individual behavior.[18] While many individuals have the ability, interest, and resources to modify behaviors independently of cultural norms and civic structure, such choices are "unlikely to bring about wider transformative change unless diffused to a broader audience that has the power to effect change through the power of numbers."[19] This is the role of law and public policy: to impact both structure and numbers and alleviate ground-level hindrances to building a new agricultural model.

III. A BRIDGE TO NEW AGRICULTURE AVENUES

"New agriculture" attempts to overcome the obstacles of the modern industrial food model in an effort to support a more sustainable food system. Direct marketing and organic food production are perhaps the most basic forms already in place that can help in the development of a more sustainable food system, and serve as a foundation for pursuing more ambitious alternative food programs.

A. Direct Marketing: Farmers' Markets, and Community-Supported Agriculture

An avenue to the new agricultural model is more direct marketing programs for farmers, such as farmers' markets and community-supported agriculture (CSA) programs.[20] Both serve prominent roles in the recent revival of community-based agriculture.[21] Both provide access to locally grown and locally processed foods, often offer organic products, and allow consumers to know or directly inquire from farmers how their food was grown, produced, and processed. In order for these direct marketing efforts to effectively promote a local organic food model, more farmers' markets and CSA programs must exist, and more people must use them.

A more sustainable food model may benefit not only environmental health, but also public health.[22] "Over the last forty years, two interrelated factors dominate the food/health argument: diminished access to healthy food and the rise of industrial food. Taken together, the two are believed to produce serious health problems, such as obesity and Type II diabetes."[23] Farmers' markets and CSAs provide access to healthy food and spurn the industrial food model. Already a number of organizations, like U.S. Department of Agriculture Agricultural Marketing Service's formation of the Farmers' Market Consortium, have recognized these health and environmental benefits and are developing initiatives to support the existence and use of farmers' markets.[24] And the USDA provides resources for farmers on its Web site to help farmers to take part in CSAs.[25]

1. The Market for Direct Markets

What share of the market do farmers' markets comprise? The short answer is that farmers' markets have seen significant growth in recent years, but their overall market share could be greater. In a 2006 report, the USDA concluded that "the U.S. farmers' market industry shows the sector continues to experience brisk growth, but that many newer farmers' markets have not yet been able to generate the sales volume enjoyed by older farmers' markets, raising questions as to whether current levels of industry growth can be sustained over time."[26] As seen in Figure 1, the number of farmers' markets nationwide has increased dramatically from 1,755 farmers' markets in 1994 to 8,284 in 2014.[27]

Unfortunately, farmers' market expansion has not yielded economic viability for the younger markets. New markets less than five years old make up one-quarter to one-third of all seasonal markets nationwide, and they have struggled to find both vendors and customers.[28] As a result, growth in the number of farmers' markets has not mirrored growth in sales. From 2000 to 2005, the average annual sales growth rate was 2.5 percent, while the number of farmers' markets grew by an astounding 43 percent.[29]

On the positive side, farmers' markets, while obviously providing local food, effectively provide a significant number of organic food options. In 2002, organic growers participated at more than four-fifths of markets studied by the USDA

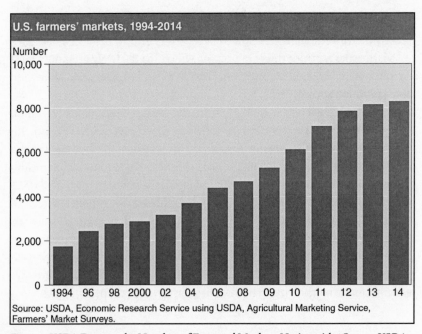

Figure 1. USDA Data on the Number of Farmers' Markets Nationwide. *Source: USDA, Economic Research Service using USDA, Agricultural Marketing Service, Farmers' Markets Surveys, http://ers.usda.gov/data-products/chart-gallery/detail.aspx?chartId=48561 &ref=collection&embed=True&widgetId=37373.*

and represented one-third of regularly attending farmers, highlighting "the disproportionately high use of farmers' markets as a sales outlet by organic growers."[30] These facts illustrate the important role of farmers' markets as a key point of purchase, providing local food and organic food at a much greater incidence than supermarkets. It also seems that farmers' markets, coupled with consumer demands, also have the ability to influence farming practices. According to one study, "customer demand for organic products has had a strong influence on some farmers who recently decided to transition to organic farming or lower-input farming practices."[31]

Like farmers' markets, CSA programs are growing in popularity. CSA is a term used to describe a group of individuals who have membership in a farm due to purchasing shares and/or volunteering on a farm. The growers/farmers and consumers/members provide mutual support and share the risks and benefits of food production of the farm.[32] Members most often buy shares in a farm in advance of the growing season to cover a farm's costs. In return, members usually receive weekly shares of farm produce, picked up at the farm or delivered to a central spot near one's residence. Other advantages include participation in farm activities, knowing one's farmer, and receipt of other raw and locally processed products such as eggs, chicken, and cheese. Members receive the benefits of farm-to-table fresh products, the bounty of great harvests, and cheaper prices. However, members also take on the risks of bad weather and a poor harvest. According to data collected by the USDA in 2007, 12,549 farms in the United States reported participating in community-supported agriculture.[33] The question is how can we both increase the number of farmers' markets and members participating in CSAs?

2. Increasing the Number and Use of Farmers' Markets and CSAs

Departments of agriculture have, in many states, been granted significant authority to encourage the development of famers' markets within the state. State agencies have grant and financial assistance programs to create new markets and improve existing ones, with a charge to expand the state's local and organic food system and support sustainable agriculture.[34] But despite important goals—"to expand the public awareness and stimulate individual use of farmers' markets to increase the purchase of locally grown foods, thereby reducing the negative environmental impact of food packaging and shipping while enhancing a beneficial economic and social climate in the community"—legislatures may only provide limited funds to achieve them.[35] States can and do provide exemptions to state labeling and packing requirements to encourage farmers' market participation.[36] However, farmers' market vendors still need to comply with health and safety standards when selling at the market,[37] which, while necessary, pose additional burdens. In addition, some farmers' markets have required vendors to provide proof of liability insurance coverage. Despite state financial assistance and regulatory changes, no models exist for a fully local and/or regional food system.

Absent a comprehensive state food system planning policy, the real player in promoting farmers' markets has been local effort, often channeled through local government. And state law often solidifies the realities of local control when it comes to farmers' markets.[38] The existence of farmers' markets is directly influenced by government policy through rules on rights of way and zoning.[39] Municipal governments play an especially important role given local control over these property rules that can "be modified to give fresh food marketing more space to sprout and security to grow."[40]

In addition to encouraging efforts to increase the total number of farmers' markets, farmers' markets must be made more economically viable. Without more and economically successful existing markets, it will prove difficult to increase the local organic share in the future. In 2004, farmers' market sales accounted for less than 2 percent of the U.S. sales overall.[41] In order to increase the economic viability of farmers' markets, two items must increase: the number of people who use farmers' markets and how much they buy. These two items are directly related. The goal of increasing these two items seem to be best met by having the markets in convenient locations, offering products in consumer demand (like more organic offerings), and developing programs that encourage regular grocery shopping at a farmers' market.

First, "[c]ustomer participation depends primarily on a market's location, since most customers tend to shop at markets close to where they live."[42] Thus, local zoning and land use regulation not only operate to offer initial existence to farmers' markets, but also their location and size, allowing markets to expand and move to better locations as demand increases and seasons change. Local zoning and ordinances can close streets, redesign parking rules, expand allowable public spaces for farmers' markets, and put markets in highly trafficked areas where people live and work. A primary reason that people do not use farmers' markets is because there is no market close to their daily life.[43] Perhaps the greatest challenge, however, to increasing consumer demand and finding a successful location is that the dominant American residential landscape is now suburban, sprawling further from downtown and filled with low-density residential developments.[44] This norm may make it difficult to put markets in a high-density residential hub except in urban neighborhoods and compact small towns with vibrant city centers.

Second, while state legislation supports marketing campaigns to increase consumer volume,[45] farmers' markets must also offer products in consumer demand. As discussed above, most markets have a significant number of organic products available, but apparently demand is so great that markets are looking for more organic farmers to join their ranks.[46] Farmers' markets are already expanding their content with processed foods, and towns are encouraging winter farmers' markets that often sell canned goods and stored root vegetables. For CSA programs, many farms are certified organic, and, to increase membership, have expanded to allow produce pickup in urban neighborhoods and town centers, in addition to the farms themselves.

Third, public policy at the state and federal level must develop programs that encourage regular grocery shopping at farmers' markets. While this goal is in part driven by location and product selection, famers' markets must also allow for a variety of payment methods and cater to customers who receive government assistance for food. Many farmers' markets are setting up credit and debit card stations to buy "market cash" that can be used.

The 2008 Farm Bill increased the commitment to federal food assistance programs by more than $10 billion over a ten-year period.[47] In 2009, the Supplemental Nutrition Assistance Program (SNAP), formerly known as the Food Stamp Program, provided over $50 million in total benefits to Americans.[48] Given these dollar figures, farmers' markets should be accessible to these program users. However, government food programs like SNAP and Women, Infants, and Children (WIC) have infrastructure that favors supermarkets, convenience stores, and other permanent indoor retailers due to electronic benefits, transfer system requirements, and government-required training.[49]

Efforts are in the initial stages to extend the benefits of these programs to farmers' markets. For example, the federal Farmers' Market Promotion Program and Senior Farmers' Market Nutrition Program are allocating 10 percent of their budgets to develop electronic benefits transfer projects at farmers' markets, support administrative costs, and provide low-income seniors with coupons that can be exchanged for eligible foods at farmers' markets, roadside stands, and community-supported agriculture programs.[50] WIC Farmers' Market Nutrition Programs, designed by Congress in 1992, have been established in many states, enabling WIC participants to use their benefits at farmers' markets.[51]

CSA programs face greater roadblocks in the face of modern economics and payment systems. CSA farms do not take credit cards, prefer payment in full at the start of the growing season in spring, and have limited methods to attract low-income customers. Data show that families with greater household income are more likely to purchase directly from farmers,[52] but CSA farms have been entrepreneurial in finding other means to bring in new customers such as sliding-scale membership fees based on income, working shares (i.e., volunteering on the farm in exchange for produce), and member share donations.

Finally, in the book *Public Produce: The New Urban Agriculture*, Darrin Nordahl argues for a greater public "market" than can be provided solely by farmers' markets. In advocating municipal agriculture, Nordahl argues that with the increased popularity of farmers' markets and community-supported agriculture, "the time is ripe to explore how we can expand this network of local food options to meet the growing demand of consumers by bringing agriculture back into our cities" through ideas such as public gardens, foraging in public places, and using vacant and government-owned space for community gardens.[53]

A theme that consistently arises in supporting sustainable agriculture is the significance of local control. Local governments influence zoning, permitting, and health and safety regulations that impact not only the approval and site of farmers' markets, but also the extent of publicly available produce. "One of the easiest ways for municipal government to support a system of public produce is to simply *allow it*."[54] For example, many municipalities have street tree ordinance bans on planting of fruit and nut trees on public streets, and sometimes local zoning prohibits small agricultural practices downtown.[55] However, the justifications of maintenance and aesthetics for not allowing food-bearing plants in public spaces may be misconceptions,[56] and localities (like Seattle and Providence) have inserted permissible language for urban agriculture through comprehensive master planning documents to effectively manage open space.[57]

B. Organic and Small Farmers

One of the most noticeable changes in the modern American food system over the last decade is the movement toward organic production and certification following passage of the Organic Food Production Act (OFPA).[58] The organic food market is flourishing, and, as a result, the modern organic production and distribution system is now dominated by large-scale "industrial organic" or "big organic" producers."[59] "The rise of commercial and industrial conventions is clear in organic distribution and consumption—where the fastest growth is in mainstream retailing, based on large-volume, regimented supply systems—and in organic production and trade—where the fastest growth is in large-scale corporate entrants pursuing organics as a high-value niche market."[60]

Organic food has almost quadrupled its market share in the last decade,[61] and organic food sales have grown from $1 billion in 1990 to over $20 billion today.[62] With large-scale production, even if organic, comes increased greenhouse gas emissions and questionable agricultural methods. For this reason, "industrial organics" have been described as "certified organic foods sold by major corporations that are technically organic but not always sustainable."[63] Organic production on small farms and in regional farming networks yields food produced and processed in a chemical-free environment that is in demand, perhaps without a large carbon footprint, and with more sustainable agricultural practices. Yet, many small farmers find it difficult to take advantage of the value-added organic label.

1. The U.S. Organic Foods Production Act and the National Organic Program

Under OFPA and the National Organic Program (NOP),[64] the U.S. government creates production, handling, and labeling standards for organic agricultural products. Individuals buy organic products to promote sustainable and chemical-free agriculture, as well as to keep their bodies free of synthetics and pesticides. OFPA establishes a national organic certification program where agricultural products may be labeled as organic if produced and handled without the use of synthetic substances.[65] The program prohibits using synthetic fertilizers, growth hormones and antibiotics in livestock, and adding synthetic ingredients during processing.[66]

Agricultural practices must follow an organic plan approved by an accredited certifying agent and the producer and handler of the product.[67] OFPA creates process-based standards but does not implement standards or require tests for actual chemical contents in food, nor assessment of overall land use practices. Thus, "certified organic" labeling informs consumers about the food production process, but does not directly describe food quality or a lack of land degradation, though organic food still is likely to have fewer chemicals than conventional counterparts.[68]

Small "farmers who gross less than $5,000 annually and only sell directly to consumers (for example, via farmers' markets and family farm stands) can avoid the certification process by simply signing a declaration of compliance" that they comply with organic standards.[69] However, if these farmers sell any of their products through conventional distribution channels, they may use the term "organic" but may not use the term "certified organic" or the USDA organic label on products without also obtaining official certification, a process that can be expensive and time-consuming.[70]

2. *The Challenges and Resources for Small Farmers*

No doubt small farmers may have trouble coming up with the funds to receive organic certification, and may also lack the resources to fully promote and market their chemical-free and sustainably grown products.[71] In recognition of the costs of organic certification for small farmers, sliding scales for payment and subsidization are the norm.[72] Organic certification fees, based on total sales, usually are below $1,000, except for large processors with far greater sales.[73] Costs are actually 75 percent less after government reimbursement if a state participates in the federal cost-share assistance program (discussed below). But perhaps due to sliding scale differences (and thus fee differences), it has been claimed that organic certifiers largely ignore issues pertaining to small-scale farmers, placing a greater emphasis on enlisting larger producers.[74]

Existing resources help with organic certification for small farmers by providing cost sharing programs. The Agricultural Management Assistance Organic

Certification Cost Share Program, established in 2001, authorizes cost share assistance to producers of organic agricultural products in a number of states and was funded $1.45 million in 2010.[75] The National Organic Certification Cost Share Program, reestablished as a part of the 2008 Farm Bill, authorizes cost share assistance to producers and handlers of organic agricultural products in each state. (Nearly every state participates.) States will reimburse each eligible producer or handler up to 75 percent of its organic certification costs, not to exceed $750.[76] In fiscal year 2008, Congress allocated on a one-time basis $22 million for this program to be allocated to states until the funds are exhausted.[77] Significant fee subsidizations exist, at least in the short term. While the costs for organic certification are expensive, they are not prohibitive, but the costs of monitoring and record-keeping may be the real barriers to entry. For example, applicants for certification must keep accurate postcertification records for five years concerning the production, harvesting, and handling of agricultural products that are to be sold as organic.[78] In addition to making organic certification more affordable for small farmers, states are also providing property tax rebates for farmers who convert from conventional to organic farming practices, and attempting to lower the tax burden on small farmers.[79]

IV. WHAT NEW AGRICULTURAL OPTIONS SHOULD BE PURSUED?

Recent legal scholarship suggests that environmental policy will focus more on individual behavior.[80] This individual behavior includes impacts of food choices on the environment.[81] In order to change modern American food consumption, two changes must take place—increased awareness and increased availability.[82] Law and policy need to (1) increase available information about the consequences of food choices, and (2) make structural changes in the food system that increase access to sustainable foods.[83] "As the availability and awareness of alternatives to industrial mass-produced food become more common, demand for something fundamentally different and better will continue to grow."[84] This is a key assertion and an argument in favor of information regulation and structural change.

A. *Increasing Information*

In his book *The Making of Environmental Law*, Professor Richard Lazarus writes, "The increased cognitive severance for consumers between environmental cause and effect exacerbates the potential environmental impact of...increased consumption."[85] This fact holds true for food consumption. Perhaps the biggest impediment to developing a more sustainable food system is a lack of food literacy. There is a large disconnect between the food we eat and knowledge of where it comes from and how it is grown.[86] Writes John Ikerd, "Nowhere in the United States is this social disconnectedness more evident than in our systems of food and farming.... Most consumers, particularly younger consumers, have no sense of where their food actually comes from or who produces it."[87]

Organic labeling alone, while a good first step, remains insufficient as the term "organic," which considers chemical input practices, does not denote sustainable because the label does not consider carbon emissions and land degradation, among other ecological concerns.[88] But information can play a useful role, especially where people are genuinely interested in a subject. The popular press and media have found success in discussing food (for example, the books of Michael Pollan), the local food movement is strong, organic food has risen in popularity, and "[t]he debate over the organic standards generated more public response than any other rule ever proposed by the USDA."[89]

In an effort to change food choices and inform consumers of the environmental impacts of food, my previous work has already argued for better informational regulation, creation of an eco-label for food-based environmental life-cycle analysis from production to use to distribution, and building on existing organic and carbon labeling programs.[90] Government-sponsored effective environmental labeling of food is an important step toward building a more sustainable food system. In addition, public "educational programs are needed to reacquaint us with food."[91] Program ideas include municipal demonstration and schoolyard gardens,[92] and food education in the primary school curriculum.[93]

Research shows that diet, especially protein sources, significantly influences environmental impacts, likely more than eating local.[94] U.S. newspaper coverage of food systems' effects on climate change has increased, but still has not reflected the increasing significant evidence of the importance of these effects.[95] This further illustrates the importance of public education about the environmental impacts of food choices, and further suggests that new dietary guidelines based on public health and the environment may be useful.[96]

B. Structural Change—Theory

As discussed above in Part II, two major structural initiatives in the food systems are already being pursued—direct marketing through farmers' markets and increased interest in organic foods. Possibilities for other structural initiatives abound. However, before addressing the merits of these possibilities, it is worth noting at the outset the significance of local control in implementing structural changes in a more sustainable food system. "It is in state legislatures and city councils, county boards, and planning commissions, and the day-to-day to-and-fro that is community life that decisions affecting our food systems are increasingly being made."[97]

Not only do localities control permitting and zoning for the likes of farmers' markets and community gardens,[98] but local communities provide the base of support and implementation for the local and slow food movements, farm-to-table, and farm-to-school. However, it is difficult to determine how state and local programs might change food systems structures in a manner that effectively improves the ecological consequences of food choice.[99] This is an empirical inquiry, made more difficult by the fact that there are so many shades of gray in environmental law as it relates to agricultural policy. Comments James E. McWilliams, author of *Just Food: Where Locavores Get It Wrong and How We Can Truly Eat Responsibly*, "[o]ur accepted dichotomies—conventional/organic, small/industrial, free range/confined, local/global, etc.—are useful in getting articles published, but they only make sense at the extremes. Most of agricultural life, however, happens between the extremes."[100]

First, environmental, agricultural, and food policy does not start from scratch. Corn is already grown in the Midwest, huge industrial farms already exist, and some climates and soils are most productive for some agricultural products, so there may be consequences like economic inefficiencies, energy waste, and new land degradation if operations are moved. Efficiency and food choices must be framed in light of the existing landscape, and scientific inquiry should dictate policy in its determination of which existing structures in the food system have the greatest and least impacts on the environment and climate change.[101]

Second, the best potential changes, or combination thereof, in food systems are not particularly clear. As some question, is the "solution for people to produce, prepare, and consume sustainably grown local food"?[102] Maybe, and maybe only in some circumstances, but not all. "Actually, networks of interdependent community-based systems in the future might serve the total food market more easily, efficiently, and effectively than can a giant, hierarchically managed, corporately controlled, and centrally planned global food chain."[103] Again, maybe, but then why does so much consolidation already exist? The findings of Weber and Matthews in *Food-Miles and the Relative Climate Impacts of Food Choices in the United States* suggest that making a dietary change would better lower an average household's food-related carbon footprint than buying local.[104] If one is skeptical about food miles, perhaps we need a combination of multiple changes. The key question, which no one answers, is an empirical one: What food systems should be promoted, to what extent, and why?

Absent this information, words of caution, or, at worst, hostility, are levied against new trends in food and agricultural policy. A recent op-ed in the *New York Times*, entitled "Math Lessons for Locavores," is illustrative.[105] The opinion piece proclaims that,

> Words like "sustainability" and "food miles" are thrown around without any clear understanding of the larger picture of energy and land use.... The result has been all kinds of absurdities. For instance, it is sinful in New York City to buy a tomato grown in a California field because of the energy spent to truck it across the country; it is virtuous to buy one in a lavishly heated greenhouse in, say, the Hudson Valley.[106]

But in an equally heated response to this op-ed, McWilliams states that we are left with a problem. "[T]here are many theoretical advantages to consolidating the food system—food can be cheaper, more accessible, more reliably diverse, and less dependent on extensive land and labor—but the underlying realities—perverse incentives, trade agreements, and subsidies—too often prevent these advantages from being realized."[107] Anna Lappé, author of *Diet for a Hot Planet*, responds in similar fashion and dismisses the "comparative advantage" argument that advocates raising crops in "places where they grow best and with the most efficient technologies."[108] She rejects this view, not because reasonable people and even locavores disagree with it, since most would agree that "choices farmers make about what foods to grow, and what time of year to grow them, should be informed by place."[109] Instead, the problem is the cause of the existing comparative advantage, those same economic and regulatory realities suggested by McWilliams. Thus, a local, organic, and less processed food system is a response to these structural barriers.

At minimum, any critique of new agricultural models reflects the importance of a more holistic food model whether by informational tools, like an eco-label and public education, or food system infrastructure, like public markets and increasing access to sustainable and organic food; local food alone is not enough. That said, local food models are clearly on the rise through community food system initiatives including farm-to-table and farm-to-school programs that create opportunities for local farms to sell directly to restaurants and schools.[110] In addition to farm-to-table, retailers and local restaurants are putting more information about their food on menus and signs, including whether the products and ingredients are organic and locally grown, and have embraced a slow food movement that spurns processed ingredients and embraces local farming. Local communities are developing locavore challenges to encourage more sustainable diets,[111] and "WWOOFing" is becoming the rage for college student summers. (WWOOF stands for Willing Workers on Organic Farms.) And state laws are in existence that encourage or mandate locally produced and locally processed food purchases for state institutions and schools.[112]

But we are still left with this open question: Given the need for a variety of structural changes to create a more holistic and sustainable food model, which efforts should be pursued initially? First, I have already argued that information devices like eco-labeling and public education are necessary to affect consumer food choices.

Second, regulatory structures must be changed as they are geared toward large agribusiness, undercutting the ability of smaller producers and processors to survive. Due to public health and safety concerns, even small-scale processing requires the use of commercial kitchens, and, except for very small-scale operations, farmers must have their animals slaughtered at USDA-approved off-site meat processing facilities.[113] Similarly, organic certification may hurt small-scale players. Writes Raynolds,

> In short, certification represents a powerful new form of network governance which is rooted in social, legal, and bureaucratic institutions, yet serves in many ways to accentuate traditional economic inequalities between firms and countries.... Powerful corporate retailers and branders also benefit from organic certification, since chain of custody and documentation requirements facilitate their participation in mainstream markets.[114]

Admittedly, modern food economics may work against small producers and processors since their size may cause them trouble with product reliability and availability, as well as keeping food costs low and predictable.[115]

Third, structural changes should reach both production and distribution channels. Processing and production account for the greatest portion of fossil fuel usage and greenhouse gases in the food system, on account of the rise of industrial foods and moving cooking out of the kitchen to the factory.[116] To many this fact suggests a limitation to the locavore movement; I would instead argue that local structures, like farmers' markets and CSAs, allow more options for consumers to buy goods that are raw, organic, and unprocessed. Similarly, food education and informational labeling would increase purchase of less-processed goods, often found on the edges of the grocery store floor. And an industrial local food market[117] may be problematic if a dominant corporation could dictate market parameters like quantities, growing conditions, and eventually put smaller farmers out of business by dictating price.[118]

But local food movements matter because transportation costs matter, and the environmental costs of food transportation will matter more over time as food distribution systems rely more heavily on air transit. Transport of food by air has the highest carbon dioxide emissions per ton and is the fastest growing mode of food transport.[119] In a German study of energy requirements for domestic apples as compared with imported New Zealand apples, the domestic apples, whose primary energy need is cold storage, required 27 percent less energy than the imported New Zealand apples, which required energy for shipping and ground transport.[120] Structural change must embrace a more holistic and sustainable food model that considers all attributes—distribution distance (i.e., food miles) and type,[121] chemical inputs (i.e., low-input, organic),[122] and level of processing.[123]

C. Structural Change—Practice

The complexities of the modern food system influence the possibilities for potential structural change within the food system. We already have an existing food system that serves as the baseline from which any change will occur. Legal regulation favors large agribusiness. And we need a more holistic food model that takes account of all phases of production and distribution, and various ideals of sustainable food (local, low-input/organic, and less processed). But the question remains, what structural changes to pursue in light of these more abstract conclusions?

Admittedly, this chapter focuses on more incremental structural changes to help individuals who usually partake in the industrial food model to be part of a more local, less processed, and more organic food system. While I recognize the need for a massive overhaul in the American food system in the interests of public and environmental health, food and national security, and an affordable *and* healthy food economy, I advocate three structural initiatives, combined with the information efforts and structural efforts already discussed (for example, eco-labels, public education, farmers' markets, CSAs, and organic certification for small farmers), to help spark this effort. These structural incentives are:

1.   Create better food system planning through state food policy councils and municipal planners.

2. Build on existing interests in intrastate and regional efforts support-
    ing local food and local economies.
3. Improve management of existing alternative agricultural distribution
    and production systems.

First, branches of government, through state legislative enactment and
gubernatorial executive orders,[124] have and should create food policy councils
(FPCs), implementing early arguments that FPCs must be formally institutionalized to be effective.[125] FPCs have been created to examine the operation of state
and local food systems, provide ideas and policy recommendations for how they
can be improved, and support food system programs.[126] FPCs at the state and
local level can also influence institutional purchases and address concerns about
food security, hunger, farmland preservation, and food labeling.[127] And farm-
to-school programs are clearly one of the most common and popular FPC and
legislative program,[128] with regional foodsheds and food hubs on the horizon.

While FPCs are a nice start in state governments, there is a lack of consideration given to food systems in local and regional planning. Unlike land use,
housing, transportation, and the environment, and more recently health, education, and energy, "[t]he food system . . . is notable by its absence from the writings
of planning scholars, from the plans prepared by planning practitioners, and from
the classrooms in which planning students are taught."[129] As I have argued earlier,
and due to the only recent interest in food systems, it is unclear what initiatives
should be pursued by FPCs and planners. In this respect, both FPCs and planners
can play a role in data acquisition and analysis to evaluate local and regional food
systems. And perhaps most importantly, FPCs may be in the best position to
implement the very recommendations of this chapter—increase visibility and
access to regional and local food; support food and nutrition programs; support
intrastate purchasing practices; and implement awareness campaigns.[130]

Second, in-state food systems and state purchasing power are perceived as
a means to improve intrastate economic conditions, and this perception can be
exploited to create a more sustainable food system. State laws are already requiring that state government and government-sponsored entities purchase locally
grown food and in-state dairy products.[131] For example, in Illinois, the new Farm-
to-School database will create an electronic database on the Department of Agriculture Web site that allows the state to connect with local farmers to purchase
fresh produce.[132] Oregon has enacted a law allowing public agencies to purchase
in-state agricultural products even if they are 10 percent more expensive than

out-of-state products.[133] These state institutional purchase programs, requiring the purchases of intrastate food products, also dovetail nicely with local identity labeling programs to promote and market in-state produce and products.[134]

Third, improvement in management and marketing of existing alternative agricultural distribution and production systems is needed to ensure that these systems do not collapse and remain viable, as well as improve access. While I already discussed above efforts to increase access to farmers' markets for lower income customers,[135] better farmers' market management is needed. While the overall number of farmers' markets is increasing, "nearly half of new markets close in the first 4 years."[136] Professional market managers are needed for marketing and strategic decisions, and government resources can be allocated for their hire.[137]

Specifically, managers are needed to help expand the consumer base (as opposed to just vendor numbers) by dealing with practical concerns like space constraints, parking, and creating better relationships with community members and local government as a way to improve promotion and funding.[138] The "farmers' market system has reached a level that demands higher levels of management, greater coordination and more effective governance."[139] Going forward farmers' market managers will have to explore permanent structures to increase consumer volume, and state and local governments, if they desire to reorient our food system, will have to play a new role in supporting local food markets through marketing and financial incentives.[140] "With farmer participation stretched thin . . . it is vital that other entities (e.g., non-governmental or governmental organizations) take lead roles in organizing and operating markets."[141]

V. CONCLUSION

There is growing interest in learning about the environmental impacts of our food choices, and in modifying individual behaviors and choices that have adverse ecological effects. "Integrating sustainable consumption and production principles into everyday patterns of behavior is a major policy challenge for governments seeking long-term sustainability, yet there is an acknowledged need for tools and instruments to put this into practice."[142]

In order to create a more sustainable food system, these tools must include information and structural change. As I've noted in my past work, law and public policy should increase available consumer information about the consequences of food choices and—the focus of this chapter—make structural changes in the food system that increase access and help form a more sustainable food system. Information, through eco-labeling and food education programs, will help play a role in changing consumer preferences. And the knowledge gained through creating these informational tools, like environmental life-cycle analysis, will help "identify the most energy-draining stages of consumption."[143]

So far, common structural avenues for promoting an organic local food system are farmers' markets, community-supported agriculture, and encouraging organic certification. Progress has been made at both the federal and state levels to find financial and technological avenues to increase producer and consumer access to these programs. Moving forward, structural change must include better food system planning, increased government support for local food and regional economies, and improved management of alternative agricultural distribution and production systems. Admittedly, these are small steps, but better information and improved structural systems to increase access to better food may shift individual norms.

Finally, I must address a final point that continues to perplex this author as well as others—price. Any local, low-input food system cannot be considered successful if locals cannot afford local food, and many individuals do not have access to healthy fruits, vegetables, and grains. To date, scholars and food policy writers (myself included) have inadequately dealt with the issue of price. For example, while I advocate creation of a food eco-labeling so people know ecological costs of processed foods,[144] already individuals cannot afford organically labeled food. "Essentially, we have a system where wealthy farmers feed the poor crap and poor farmers feed the wealthy high-quality food."[145] I take this to mean that agribusiness, supported by governmental policy, offers up processed industrial foods with commodity grains, and small rural farmers have moved to organic produce and artisanal processed goods.[146] This illustrates the importance of implementing policy that allows low-income individuals to use public assistance programs at farmers' markets and other similar programs.[147]

One certainly sees that high-calorie mass-produced foods are increasing in price at lower rates than healthier foods, and many healthy foods (for example, good produce) are not available in many poor urban neighborhoods. Both in the United States and abroad, the "current pricing system externalizes social and environmental costs and benefits, and this, together with current subsidy systems for intensive pesticide-dependent monoculture, results in local organic produce costing more than conventionally grown imported food."[148]

In addition, it is true that having an organic locavore diet is becoming a sign of being of higher socioeconomic status, and, thus, we must not lose sight of social justice concerns like hunger and food insecurity.[149] I do not believe, however, we should undervalue the importance of information, as discussed, and food literacy (for example, knowing where your food comes from, how to cook, and what is healthy), or underestimate the power of marketing for unhealthy industrial food. What remains, however, is a challenge for making the economics of a sustainable food system work, and understanding the value of healthy and ecologically sound food choices. How can society afford healthy local low-input food, and why are we spending so much less of our income on food?

NOTES

1. *See, e.g.*, Mary Jane Angelo, *Corn, Carbon, and Conservation: Rethinking U.S. Agricultural Policy in a Changing Global Environment*, 17 Geo. Mason L. Rev. 593 (2010); Jason J. Czarnezki, *The Future of Food Eco-Labeling: Organic, Carbon Footprint, and Environmental Life-Cycle Analysis*, 30 Stan. Envtl. L.J. 3 (2011); William S. Eubanks II, *A Rotten System: Subsidizing Environmental Degradation and Poor Public Health with Our Nation's Tax Dollars*, 28 Stan. Envtl. L. J. 213 (2009).

2. "Food miles" is the term used to describe the distance food is transported from farm to table.

3. This chapter focuses on the ecological costs of food. *Cf.* Patricia Allen and Martin Kovach, *The Capitalist Composition of Organic: The Potential of Markets in Fulfilling the Promise of Organic Agriculture*, 17 Agric. & Hum. Values 221, 221 (2000) ("We have chosen to focus on environmental issues because improving environmental conditions in agricultural production is the most significant and consistent claim made by advocates of organic agriculture"). But food matters in other senses as well. Alice Waters, *Foreword*, in Carlo Petrini, Slow Food Nation, at ix (2007) ("We soon discovered that the best tasting food came from local farmers, ranchers, and foragers, and fisherman who were committed to sound and sustainable practices.").

4. Elisabeth Rosenthal, *To Cut Global Warming, Swedes Study Their Plates*, N.Y. Times, Oct. 23, 2009, at A6.

5. *See* Czarnezki, *supra* note 1. This would go beyond organic labeling under the OFPA. It would also go beyond regional food labeling. *See, e.g.*, Amy B. Trubek and Sarah Bowen, *Creating the Taste of Place in the United States: Can We Learn from the French?* 73 GeoJournal 23 (2008).

6. Jill Richardson, Recipe for America: Why Our Food System Is Broken and What We Can Do to Fix It 47 (2009) (in terms of ecological interests, defines sustainable agriculture as "resource-conserving," "environmentally sound," and efficient use of resources); Gill Seyfang, *Ecological Citizenship and Sustainable Consumption: Examining Local Organic Food Networks*, 22 J. Rural Stud. 383, 383 (2006) (recognizing that sustainable consumption has an elusive definition).

7. Michael Pollan, *Farmer in Chief*, N.Y. Times Mag., Oct. 12, 2008, at MM62.

8. William S. Eubanks II, *Paying the Farm Bill: How One Statute Has Radically Degraded the Natural Environment and How a Newfound Emphasis on Sustainability Is the Key to Reviving the Ecosystem*, 27 Envtl. Forum 56 (2010).

9. Wynne Wright and Gerad Middendorf, *Fighting over Food: Change in the Agrifood System*, in The Fight over Food: Producers, Consumers, & Activists Challenge the Global Food System 7–8 (Wynne Wright and Gerad Middendorf, eds., 2008).

10. Thomas A. Lyson, Civic Agriculture: Reconnecting Farm, Food, and Community 63 (2004).

11. Wright and Middendorf, *supra* note 9, at 2.

12. Neil D. Hamilton, *Putting a Face on Our Food: How State and Local Food Policies Can Promote the New Agriculture*, 7 Drake J. Agric. L. 407 (2002).

13. *Cf.* Darrin Nordahl, Public Produce: The New Urban Agriculture, at xii (2009) (suggesting a need for a "public network of food-growing opportunities" due to rising cost of produce, weather aberrations and subsequent crop loss, pathogen-infected produce, decreasing popularity of industrial organic, and demand for locally grown produce).

14. *See* Wright and Middendorf, *supra* note 9, at 1.

15. John E. Ikerd, Crisis & Opportunity: Sustainability in American Agriculture 293 (2008) ("The industrial era has been fueled by *cheap* energy," specifically fossil fuels [emphasis in original]).

16. Wright and Middendorf, *supra* note 9, at 9 ("Local food systems are 'rooted in particular places, aim to be economically viable for farmers and consumers, use ecologically sound production and distribution practices and enhance social equity and democracy for all members of the community'" [citations omitted]).

17. *See* Hamilton, *supra* note 12.

18. *See generally* Jason J. Czarnezki, Everyday Environmentalism: Law, Nature & Individual Behavior (2011).

19. Wright and Middendorf, *supra* note 9, at 15. *See also* Laura B. DeLind, *Are Local Food and the Local Food Movement Taking Us Where We Want to Go? Or Are We Hitching Our Wagons to the Wrong Stars*, Agric. Hum. Values (Feb. 22, 2010) (on file with author) (arguing that advocating individual action can deflect responsibility and can starve social or political activism).

20. Raymond A. Jussaume Jr. and Kazumi Kondoh, *Possibilities for Revitalizing Local Agriculture: Evidence from Four Counties in Washington State*, in The Fight over Food, *supra* note 9, at 236.

21. Wright and Middendorf, *supra* note 9, at 9.

22. Nick Rose et al., *The 100-Mile Diet: A Community Approach to Promote Sustainable Food Systems Impacts Dietary Quality*, 3 J. Hunger & Envtl. Nutrition 270, 282 (2008) (suggesting further research is needed on the relationship between sustainable food diet and health effects).

23. Alfonso Morales and Gregg Kettles, *Healthy Food Outside: Farmers' Markets, Taco Trucks, and Sidewalk Fruit Vendors*, 26 J. Contemp. Health L. & Pol'y 20, 30 (2009) (citing Centers for Disease Control, *The Burden of Chronic Diseases and Their Risk Factors: National and State Perspectives 2004* 29, 44 (2004), *available at* http://9healthfair.publichealthpractice.org/documents/burden_book2004.pdf).

24. *See* USDA, Agricultural Marketing Service, *Farmers Markets and Local Food Marketing*, USDA, Agricultural Marketing Service, http://www.ams.usda.gov/AMSv1.0/Farmers

Markets; Farmers Market Coalition *About FMC,* Farmers Market Coalition, http://farmersmarketcoalition.org/joinus/.

25. USDA, *CSA Resources for Farmers,* National Agricultural Library, http://www.nal.usda.gov/afsic/pubs/csa/csafarmer.shtml.

26. Edward Ragland and Debra Tropp, USDA, National Farmers Market Manager Survey 2006 1 (2006).

27. *See also* Morales and Kettles, *supra* note 23, at 27–28 (discussing the history and number of public markets and street vendors in the United States).

28. Ragland and Tropp, *supra* note 26. ("As a result of the massive expansion in the number of farmers markets since 2000, nearly 30 percent of all seasonal markets are less than 5 years old and most still appear to be establishing themselves economically. Managers of these young markets reported monthly sales only half the national average of all markets. They also reported fewer vendors [22 compared with a national average of 31] and fewer customers per week [430 compared with a national average of 959]").

29. *Id.* (stating that "the large percentage of young markets explains in part why the growth in the number of farmers markets is not mirrored by a corresponding growth in sales. Total farmers market sales in 2005 are estimated to have slightly exceeded $1 billion, compared with $888 million in 2000, an average annual growth rate of 2.5 percent," and noting the 43 percent growth in the number of farmers' markets between 2000 and 2005).

30. Amy Kremen, Catherine Greene, and Jim Hanson, *Organic Produce, Price Premiums, and Eco-Labeling in U.S. Farmers' Markets* 4 (Economic Research Service, USDA 2004), *available at* http://www.ers.usda.gov/publications/VGS/Apr04/vgs30101/vgs30101.pdf.

31. *Id.* at 6.

32. USDA *Community Supported Agriculture,* National Agricultural Library, http://www.nal.usda.gov/afsic/pubs/csa/csa.shtml; Suzanne DeMuth, *Community Supported Agriculture (CSA): An Annotated Bibliography and Resource Guide,* excerpted in USDA *Defining Community Supported Agriculture,* National Agricultural Library (Sept. 1993), http://www.nal.usda.gov/afsic/pubs/csa/csadef.shtml; *see also* Wright and Middendorf, *supra* note 9, at 10.

33. USDA, *National Agricultural Statistics Service,* 2007 Census of Agriculture—United States Summary and State Data 49 tbl. 44 (2009), *available at* http://www.agcensus.usda.gov/Publications/2007/Full_Report/usv1.pdf.

34. *See, e.g.,* Conn. Gen. Stat. § 22-6j, k (2010) (state grant program to set up farmers' markets); N.Y. Agric. & Markets Law (Consol. 2011) § 262 et seq. (2010) (state assistance for farmers' markets including construction, reconstruction, improvement, expansion, and rehabilitation as well as promotional support); Pa. Cons. Stat. § 2403 (2010) (helping to develop farmers' market business plans, predevelopment costs, promotions, marketing, management operation); 10 Vt. Stat. Ann. § 330 (West 2010) (supporting farm-to-plate investment program, farmers' markets, and CSAs); Ga. Code. Ann. § 2-10-57 (2010) and Minn. Stat. § 17.114 (2009) (giving state agencies authority to regulate and promote farmers' markets and support sustainable agriculture); 505 Ill. Comp. Stat. 84/15 (2011) (Illinois Local and Organic Food and Farm Plan with the goal to expand farmers' markets and state local and organic food system).

35. Ark. Code. Ann. § 20-83-102 (2010).

36. *See* Cal. Food & Agric. Code § 47002 (Deering 2010).

37. *See, e.g.,* Cal. Health & Safety Code § 114371 (Deering 2009).

38. *See, e.g.,* Cal. Food & Agric. Code § 47004(b) (Deering 2009) ("Certified farmers' markets are locations established in accordance with local ordinances . . ."); Nev. Rev. Stat. Ann § 268.092 [Lexis-Nexis 2009]).

39. Morales and Kettles, *supra* note 23, at 40–41.

40. *Id.* at 22.

41. Kremen, Greene, and Hanson, *supra* note 30, at 2.

42. *Id.* at 2. *See also* Richardson, *supra* note 6, at 78 (stating that "the best markets are located in a convenient, central place, with ample parking and perhaps even bike racks," have good hours, and continue through the winter months).

43. *See, e.g.,* Ramu Govindasamy et al., *Farmers' Markets: Consumer Trends, Preferences, and Characteristics* 4 (1998), http://njveg.rutgers.edu/assets/pdfs/mktg/fm_consumer_trends _june1998.pdf.

44. Dolores Hayden, Building Suburbia: Green Fields and Urban Growth, 1820–2000, at 3 (2003); *see generally* Kenneth T. Jackson, Crabgrass Frontier: The Suburbanization of the United States (1987).

45. *See, e.g.,* Conn Gen. Stat. § 22-38a (2010); 3 Pa. Cons. Stat. § 2403 (2010).

46. Kremen, Greene, and Hanson, *supra* note 30, at 11.

47. USDA, Food and Nutrition Service, *A Short History of SNAP*, Supplemental Nutrition Assistance Program (SNAP), http://www.fns.usda.gov/snap/rules/Legislation/about.htm.

48. USDA, *National Level Annual Summary: Participation and Costs*, Supplemental Nutrition Assistance Program Participation (SNAP) (Jan. 31, 2011), *available at* http://www.fns.usda .gov/pd/SNAPsummary.htm.

49. Morales and Kettles, *supra* note 23, at 22–23.

50. USDA, Agricultural Marketing Service, *2009 Farmers Market Program Promotion Guidelines* 9 (2009), *available at* http://www.ams.usda.gov/AMSv1.0/getfile?dDocName= STELPRDC5075760; USDA, Food and Nutrition Service, *Overview*, Senior Farmers' Market Nutrition Program, http://www.fns.usda.gov/wic/SeniorFMNP/SeniorFMNPoverview .htm.

51. Morales and Kettles, *supra* note 23, at 40–41 (citing Albemarle State Policy Center, *Balance: A Report on State Action to Promote Nutrition, Increase Physical Activity and Prevent Obesity,* [2007], at 107, *available at* http://www.policyarchive.org/handle/10207/21478); *See also* California Department of Food and Agriculture (CDFA), Certified Farmers' Market Advisory Committee (CFMAC), Meeting Minutes (Oct. 29, 2008), *available at* www.cdfa .ca.gov/is/docs/CFMAC_Minutes102908.pdf; Neil Hamilton, The Legal Guide for Direct Farm Marketing 39 (1999); USDA, Food and Nutrition Service, WIC Farmers' Market Nutrition Program, http://www.fns.usda.gov/wic/FMNP/FMNPfaqs.htm#1. Many states have farmers' market nutrition programs that implement the federal program (WIC Famers' Market Nutrition Act of 1992); *see, e.g.,* Ariz. Rev. Stat. §36-700 (2010); Cal. Health & Safety Code § 123279 (2010); Conn Gen. Stat. §§ 22-6g to 6q (2010); Iowa Code § 175B (2010); Ky. Rev. Stat. § 260.031 (2010).

52. Raymond A. Jussaume Jr. and Kazumi Kondoh, *Possibilities for Revitalizing Local Agriculture: Evidence from Four Counties in Washington State*, in The Fight over Food, *supra* note 9, at 239 (2008) (in a study of Washington counties, showing the "greater the household income, the more likely an individual is to shop directly from a farmer").

53. Nordahl, *supra* note 13, at 45–46.

54. *Id.* at 53 (emphasis in original).

55. *Id.* at 54, 56, 133 ("Municipalities should encourage, rather than forbid, home and business owners to plant edibles in the right-of-way").

56. *Id.* at 91.

57. *Id.* at 57–58.

58. Organic Foods Production Act of 1990, 7 U.S.C. § 6501 (1990).

59. Kate L. Harrison, *Organic Plus: Regulating Beyond the Current Organic Standards*, 25 Pace Envtl. L. Rev. 211, 212 (2008) (citing James Temple, *The 'O' Word: Some Organic Farmers Opt Out of Federal System*, Contra Costa Times, Oct. 29, 2006, at 6B; *Earthbound Farm Gains Efficiencies with Supply Chain Execution Solutions from HighJump Software*, Business Wire, June 14, 2005, at 1).

60. Laura T. Raynolds, *The Globalization of Organic Agro-Food Networks*, 32 World Dev. 725, 738 (2004).

61. Organic Trade Ass'n, *Executive Summary: Organic Trade Association's 2007 Manufacturer Survey* (2007), *available at* http://www.ota.com/pics/documents/2007ExecutiveSummary .pdf (stating that organic food sales accounted for 0.8 percent of total food sales in 1997, and 2.8 percent in 2006).

62. Organic Trade Ass'n, *The Organic Industry*, *available at* http://www.ota.com/pics /documents/Mini%20fact%201-08%20confirming.pdf.

63. Richardson, *supra* note 6, at 11.

64. NOP is organized under the Organic Foods Production Act of 1990 and under 7 C.F.R. pt. 205. See USDA, Agricultural Marketing Service *National Organic Program*, http://www .ams.usda.gov/AMSv1.0/nop.

65. Organic Foods Production Act of 1990, 7 U.S.C. § 6501 (1990).

66. *Id.* §§ 6508(b)(1), 6509(c)(3), 6510.

67. *Id.* §§ 6504–6505.

68. Michelle T. Friedland, *You Call That Organic?—The USDA's Misleading Food Regulations*, 13 N.Y.U. Envtl. L. J. 379, 398–99 (2005). However, "because food produced in accordance with the NOP regulations will not be intentionally sprayed with pesticides or intentionally grown or raised using genetically engineered seed or other inputs, the likelihood of the presence of pesticide residue or genetically engineered content will clearly be lower than in foods intentionally produced with pesticides and genetic engineering techniques. But organic food will not be free of such contamination. Evidence clearly indicates that both pesticides and genetically engineered plant materials often drift beyond their intended applications, and organic food, like any food, may be accidentally contaminated." *Id.* at 399–400.

69. Harrison, supra note 59, at 219 (citing Andrew J. Nicholas, As the Organic Industry Gets Its House in Order, the Time Has Come for National Standards on Genetically Modified Foods, 15 Loy. Consumer L. Rev. 277, 285 (2003).

70. *Id.*

71. Richardson, *supra* note 6, at 63–64 ("Because it costs money and takes time to achieve organic certification, some farmers choose not to get certified, even if they may meet or exceed USDA organic standards").

72. *See generally* Ariana R. Levinson, *Lawyers as Problem-Solvers, One Meal at a Time: A Review of Barbara Kingsolver's Animal, Vegetable, Miracle,* 15 Widener L. Rev. 289 (2009).

73. *See, e.g., Vermont Organic Farmers, LLC—Timeline and Fees for Certification,* Northeast Organic Farming Association of Vermont, http://nofavt.org/programs/organic-certification /application-deadline-and-fees; Oregon Tilth, *Certified Organic Fee Schedule* (2011), *available at* http://tilth.org/files/certification/OTCOFeeSchedule.pdf.

74. Denis A. O'Connell, *Shade-Grown Coffee Plantations in Northern Latin America: A Refuge for More Than Just Birds & Biodiversity,* 22 UCLA J. Envtl. L. & Pol'y 131 (2003/2004) (citing Russell Greenberg, *Criteria Working Group Thought Paper* 4 [2001]).

75. USDA Agricultural Marketing Service (AMS), *National Organic Program Cost Share Programs 2010 Report to Congress* 1, *available at* http://www.ams.usda.gov/AMSv1.0/getfile ?dDocName=STELPRDC5084541&acct=nopgeninfo (hereinafter USDA AMS).

76. Press Release, USDA, *USDA Amends National Organic Certification Cost Assistance Program* (Nov. 7, 2008), *available at* http://www.usda.gov/wps/portal/usda /usdahome?contentid=2008/11/0288.xml.

77. USDA AMS, *supra* note 75. To prevent duplicate assistance payments, producers participating in the AMS program are not eligible to participate in the producer portion of the national program.

78. USDA Agricultural Marketing Service, *National Organic Program: Certification* (Apr. 2008), *available at* http://www.ams.usda.gov/AMSv1.0/getfile?dDocName =STELDEV3004346&acct=nopgeninfo.

79. *See, e.g.,* Ga. Code Ann. § 48-5-41 (2006); Woodbury County, Iowa, Resolution: Organics Conversion Policy, (June 2005), *available at* http://www.farmlandinfo.org/woodbury -county-ia-organics-conversion-policy.

80. *See, e.g.,* Czarnezki, *supra* note 18; Hope M. Babcock, *Assuming Personal Responsibility for Improving the Environment: Moving Towards a New Environmental Norm,* 33 Harv. Envtl. L. Rev. 117 (2009); Michael Vandenbergh, *The Individual as Polluter,* 35 ELR 10723 (2005); Michael P. Vandenbergh, *Order without Social Norms: How Personal Norm Activation Can Protect the Environment,* 99 N.W. Univ. L. Rev. 1101 (2005).

81. *See* Czarnezki, *supra* note 18.

82. In addition to the two topics addressed in this chapter, other changes must occur. As stated earlier, massive overhaul of the Farm Bill is likely necessary. And a final prong, which I haven't written about to this point, will involve farmers, farming, and actual agricultural practices. *Cf.* Ikerd, *supra* note 15, at 284 ("They [farmers] are rediscovering the fundamental roots of agriculture; they are reconnecting to the land and to each other; and in the process are redefining farming."). For a discussion of many potential characteristics of a sustainable

food system, *see* Jack Kloppenburg Jr. et al., *Tasting Food, Tasting Sustainability: Defining the Attributes of an Alternative Food System with Competent, Ordinary People,* 59 Hum. Org. 177 (2000). However, some research suggests that current sustainable farming practices may not be driven by environmental concerns, but instead by an interest in preserving farmland and local family farms. Theresa Selfa et al., *Envisioning Agricultural Sustainability from Field to Plate: Comparing Producer and Consumer Attitudes and Practices Toward "Environmentally Friendly" Food and Farming in Washington State, USA,* 24 J. Rural Stud. 262, 273–74 (2008).

83. *See* James E. McWilliams, *Why Can't We All Just Sit Down and Eat Nicely Together?,* posting to *Food Fight: Do Locavores Really Need Math Lessons?,* Grist, Aug. 25, 2010, http://www.grist.org/article/food-fight-do-locavores-really-need-math-lessons/ (noting that "not everyone has the choice to opt out and hit the farmers market. For many reasons, local food choices aren't a reality for most consumers," thus illustrating the need for structural changes).

84. Ikerd, *supra* note 15, at 288.

85. Richard J. Lazarus, The Making of Environmental Law 220 (2004).

86. Nordahl, *supra* note 13, at 11.

87. Ikerd, *supra* note 15, at 277.

88. Richardson, *supra* note 6, at 11 (stating that "sustainable always means organic, but organic does not always mean sustainable"); Allen & Kovach, *supra* note 3, at 230 ("Organic labeling is simply not enough to create an agrifood system that provides real value").

89. Allen & Kovach, *supra* note 3, at 229.

90. Czarnezki, *supra* note 1.

91. Nordahl, *supra* note 13, at 11.

92. Richardson, *supra* note 6, at 89 (advocating school gardens as a way to educate kids about food).

93. Nordahl, *supra* note 13, at 118, 129–31.

94. Christopher L. Weber and H. Scott Matthews, *Food-Miles and the Relative Climate Impacts of Food Choices in the United States,* 42 Envtl. Sci. & Tech. 3508, 3512 (2008); James E. McWilliams, Just Food: Where Locavores Get It Wrong and How We Can Truly Eat Responsibly 118–19 (2010) (discussing the high environmental costs of meat consumption, and making the bold statement that "every environmental problem related to contemporary agriculture that I've investigated ends up having its deepest roots in meat production. Monocropping, excessive applications of nitrogen fertilizer, addiction to insecticides, rainforest depletion, land degradation, topsoil runoff, declining water supplies, even global warming—all these problems would be considerably less severe if global consumers treated meat like caviar, that is, as something to be eaten rarely, if ever").

95. Roni A. Neff, Iris L. Chan, and Katherine Clegg Smith, *Yesterday's Dinner, Tomorrow's Weather, Today's News? US Newspaper Coverage of Food System Contributions to Climate Change,* Pub. Health Nutrition (2008).

96. Czarnezki, *supra* note 1, at 13 (discussing Sweden's new dietary guidelines, which consider both environmental and public health).

97. Mark Winne, Closing the Food Gap: Resetting the Table in the Land of Plenty 167–69 (2009).

98. *See, e.g.*, Shady Cove Or. Code § 110.08(F), Growers' Market in Commercial Zones (1997); Annemarie Mannion, *Green Acres in the Big City: Increase in Urban Agriculture Leads to New Ordinances*, Am. City and County, July 1, 2009, http://americancityandcounty.com/admin/urban-agriculture-ordinances-200907/ (discussing municipal efforts such as Miami's new zoning ordinances that include laws regulating community gardens, rooftop gardens, greenhouses, and backyard gardens, and Milwaukee's efforts to lease five vacant lots in a central city neighborhood for use as community gardens).

99. *Cf.* Mary Story et al., *Creating Healthy Food and Eating Environments: Policy and Environmental Approaches*, 29 Ann. Rev. Pub. Health 253 (2007).

100. McWilliams, *supra* note 83.

101. In addition, while these environmental considerations matter, we cannot lose sight of other concerns such as hunger and class difference. Hunger and starvation may provide persuasive rationale for the industrial food model in some contexts. Winne, *supra* note 97, at 125–33. Winne's chapter subtitle says it all—"The Poor Get Diabetes; The Rich Get Local and Organic."

102. Erik Phillips-Nania, *Local Food Currency: An Economic Tool for Community Health*, 12 Vt. J. Envtl. L. (2011).

103. Ikerd, *supra* note 15, at 289.

104. Weber and Matthews, *supra* note 94. Buying locally will reduce greenhouse gas emissions by 4 to 5 percent, but shifting one day per week of protein from meat or dairy to vegetables, or even another protein source (fish, chicken, eggs) has the same effect as buying all household food from local providers. A *completely* local diet saves the equivalent of 1,000 miles per year driven, but a *one day per week* protein shift from red meat to chicken, fish, or eggs saves 760 miles per year, and a *one day per week* shift to veggies saves 1,160 miles per year. *Id.* at 3512–13.

105. Stephen Budiansky, Op-Ed, *Math Lessons for Locavores*, N.Y. Times, Aug. 19, 2010, at A21.

106. *Id.*

107. McWilliams, *supra* note 83.

108. Anna Lappé, *The Real Problems Locavores Are Wrestling With*, posting to *Food Fight: Do Locavores Really need Math Lessons?*, Grist (Aug. 22, 2010), http://www.grist.org/article/food-fight-do-locavores-really-need-math-lessons/P4#lappe.

109. *Id.*

110. *See Community Supported Agriculture*, USDA, National Agricultural Library, http://www.nal.usda.gov/afsic/pubs/csa/csa.shtml.

111. Angelo, *supra* note 1, at n. 6. ("The term 'locavore,' coined by Jessica Prentice on the occasion of World Environment Day 2005, describes a person who eats food grown or produced locally or within a prescribed distance. The locavore movement promotes the practice of eating locally produced food and purchasing food from farmers' markets because buying locally grown food is less energy intensive and more environmentally friendly than purchasing food from large centralized supermarkets.")

112. Hamilton, *supra* note 12, at 425–27 (citing legislation in Minnesota and California); *see also* Mass. Gen. Laws ch. 69 § 6A(b) (2010).

113. *Cf.* Richardson, *supra* note 6, at 98.

114. Raynolds, *supra* note 60, at 738.

115. *Cf.* Richardson, *supra* note 6, at 93–94.

116. McWilliams, *supra* note 94, at 25 (stating that "production and processing account for 45.6 percent of the fossil fuel usage").

117. Stephanie Clifford, *Wal-Mart Plans to Buy More Local Produce*, N.Y. Times, Oct. 14, 2010, at B1.

118. DeLind, *supra* note 19.

119. Dep't for Env't, Food, and Rural Affairs, The Validity of Food Miles as an Indicator of Sustainable Development ii (2005) (U.K.).

120. Michael M. Blanke and Bernard Burdick, *Food (miles) for Thought: Energy Balance for Locally-Grown Versus Imported Apple Fruit*, 12 Envtl. Sci. & Pollution Res. 125 (2005).

121. Andrew Martin, *If It's Fresh and Local, Is It Always Greener?*, N.Y. Times, Dec. 9, 2007, at B11.

122. Dick Cobb et al., *Integrating the Environmental and Economic Consequences of Converting to Organic Agriculture: Evidence from a Case Study*, 16 Land Use Pol'y 207, 207 (1999) ("The study showed that there were demonstrable differences in overall environmental conditions in the comparison of organic and non-organic farming, with field evidence of increased species diversity, and an eventual improvement in the profitability of the organic farming regime."); see also D.G. Hole et al., *Does Organic Farming Benefit Biodiversity?*, 122 Biolog. Conserv. 113, 123 (2005) (arguing that organic farming can play a significant role in increasing biodiversity).

123. Seyfang, *supra* note 6, at 386 (citing Jules Pretty, *Some Benefits and Drawbacks for Local Food Systems* (2001), *available at* http://www.sustainweb.org/pdf/afn_m1_p2.pdf) (finding "that environmental externalities add 3.0% to the cost of local-organic food, and 16.3% to the cost of conventional-global food").

124. Governor of Iowa, Exec. Order No. 16, 22 Iowa Admin. Bull. 1550, Apr. 19, 2000 (creating Iowa Food Policy Council); Governor of Ohio, Exec. Order 2007–27S, Aug. 7, 2007 (creating the Ohio Food Policy Advisory Council). Food policy councils can exist at the state and municipal level. *See, e.g.*, San Francisco Food Systems, http://www.sffoodsystems.org; City of Hartford Advisory Commission on Food Policy, City of Hartford, http://www.hartfordfood.org/programs/food-policy-and-advocacy/; Connecticut Food Policy Council, http://www.foodpc.state.ct.us;. For a full list of food policy councils *see* Drake University Law School, Agricultural Law Center, Food Projects and Policy, http://www.law.drake.edu/clinicsCenters/agLaw/?pageID=agFoodPolicy.

125. Kenneth A. Dahlberg, *Food Policy Councils: The Experience of Five Cities and One County* (June 11, 1994), *available at* http://unix.cc.wmich.edu/~dahlberg/F4.pdf.

126. Hamilton, *supra* note 12, at 420; Alethea Harper et al., Food Policy Councils: Lessons Learned 2 (2009).

127. Hamilton, *supra* note 12, at 420.

128. Nearly two dozen states have legislation for farm-to-school programs, as do individual school districts and municipalities. *See generally* Harper et al., *supra* note 126.

129.   Kameshwari Pothukuchi and Jerome L. Kaufman, *The Food System: A Stranger to the Planning Field*, 66 J. Am. Planning Assoc. 113 (2000), *available at* http://www.cityfarmer .org/foodplan.html.

130.   *Cf.* Portland-Multnomah Food Policy Council, *Food Policy Recommendations* (2003), available at http://web.multco.us/sites/default/files/sustainability/documents/fpc_2003 _full_report.pdf.

131.   *See, e.g.*, Vt. Act No. 38 (H. 522) (2007–08) (developing a system of local food and dairy purchasing within state government and government-sponsored entities). *See also* Brannon P. Denning, Samantha Graff, and Heather Wooten, *Laws to Require Purchase of Locally Grown Food and Constitutional Limits on State and Local Government: Suggestions for Policymakers and Advocates*, 1 J. Agric., Food Sys., & Cmty. Dev. 139, 146 (2010) (suggesting use of the "market-participant exception" to the dormant commerce clause to allow for government's direct local food purchasing or agreements with local food service contractors).

132.   Ted Gregory, *Quinn Signs Laws Promoting Local Food*, Chi. Trib., Jul. 17, 2010, *available at* http://articles.chicagotribune.com/2010-07-17/news/ct-met-farmers-market-20100717_1 _link-cards-food-stamp-recipients-john-sondgeroth.

133.   Derrick Braaten and Marne Coit, *Legal Issues in Local Food Systems*, 15 Drake J. Agric. L. 9, 32 (2010) (citing H.B. 2763, 75th Legis. Assem., Reg. Sess. (Or. 2009)).

134.   *See, e.g.*, Minn. Stat. Ann. § 17.102 (2010) (governing "Minnesota Grown" labeling licenses).

135.   Selfa, *supra* note 82, at 274 ("Thus, scholars and activists with an interest in promoting sustainable food networks could focus on facilitating ways for producers and consumers to engage in actions that reflect their interests in consuming sustainably produced food and preserving farmland and local ecologies. Perhaps more emphasis could be given to developing and expanding programs, such as WIC and senior farmers' markets nutrition programs, which provide greater access to sustainably produced food for lower income consumers").

136.   Garry Stephenson, Larry Lev, and Linda Brewer, Understanding the Link Between Famers' Market Size and Management Organization 15 (2007) (Oregon State Extension Service, Special Report No. 1082-E), *available at* http://smallfarms.oregonstate.edu/sites /default/files/small-farms-tech-report/eesc_1082-e.pdf.

137.   *See id.*

138.   Lydia Oberholtzer and Shelly Grow, *Producer-Only Farmers' Markets in the Mid-Atlantic Region: A Survey of Market Managers*, 6 (2003) *available at* https://www.researchgate .net/publication/237111558_A_Survey_of_Market_Managers.

139.   *Id.* at 20.

140.   Carolyn Dimitri, Edward C. Jaenicke, and Lydia Oberholtzer, *Local Marketing of Organic Food by Certified Organic Processors, Manufacturers, and Distributors*, 26 J. Agribusiness 157 (2008) (successful strategies for improving the food system include "local governments' promotion of local organic marketing, such as supporting local farmers markets, restaurants that rely on local products, and the sales of locally grown food in supermarkets.... A different approach might be for governments to provide incentives for retailers to carry locally grown food").

141. Oberholtzer and Grow, *supra* note 140, at 19.

142. Seyfang, *supra* note 6, at 383.

143. McWilliams, *supra* note 94, at 24.

144. Czarnezki, *supra* note 1, at 2.

145. Lisa Miller, *Divided We Eat*, Newsweek, Nov. 22, 2010, http://www.newsweek.com/2010/11/22/what-food-says-about-class-in-america.html (quoting Michael Pollan). *See also* Thomas Macias, *Working Toward a Just, Equitable, and Local Food System: The Social Impact of Community Based Agriculture*, 80 Soc. Sci. Q. 1086, 1088 (2008) ("Without a program to promote access to and knowledge about healthy food for the general public, there is a good chance food quality will be satisfied with the relatively well-off having the best access and the rest of society left with food created primarily for mass production and easy distribution, product quality being a secondary concern"); Adam Drewnowski and S. E. Specter, *Poverty and Obesity: The Role of Energy Density and Energy Costs*, 79 Am. J. Clin. Nutr. 6 (2004).

146. Ben Hewitt, The Town That Food Saved: How One Community Found Vitality in Local Food 89 (2009) (addressing concerns that local food is becoming a "gentrified green, boutique scene"). *See also* Michael Pollan, *Why Eating Well Is "Elitist,"* N.Y. Times, May 11, 2006, http://pollan.blogs.nytimes.com/2006/05/11/why-eating-well-is-elitist/ ("It is very simply a function of government policy: our farm policies subsidize the most energy-dense and least healthy calories in the supermarket.").

147. *See* Macias, *supra* note 145, at 1096, 1098.

148. Seyfang, *supra* note 6, at 90.

149. *See* Mark Winne, *Local, Organic Food for Every Budget*, Hartford Courant, Aug. 18, 2003, at A7 ("[b]ut the hot pursuit of local, organic produce stands in sharp contrast to the growth in food insecurity and hunger").

10 Breaking Our Chemical Addiction

A Twelve-Step Program for Getting Off the Pesticide Treadmill

Mary Jane Angelo, University of Florida

S ince the mid-twentieth century, the developed world, including the United States, has relied heavily on the use of synthetic chemical pesticides to support industrialized high-yield agriculture. Chemical pesticides, generally derived from fossil fuels, comprise a significant component of most industrialized monoculture agriculture. This dependency on synthetic chemical pesticides is attributable to a number of factors including ill-conceived policy choices, perverse economic incentives, and ecological reality. The combination of these factors has created a dependency that does not necessarily make economic sense, that contributes to significant human health and environmental harms, and that is counterproductive in that it can lead to ever greater pesticide dependency. The need to continue dependency on synthetic chemicals in the face of the serious environmental and social problems associated with them is, in essence, an addiction. Webster's dictionary defines the word "addiction" broadly as a "persistent compulsive use of a substance known by the user to be harmful."[1] By any account, the persistent use of pesticides, known to be harmful to human health and the environment, fits this broad definition. Even a more specific medical definition of the word "addiction," focused on the biological need for and increased tolerance of the substance, could be applied to pesticide use.[2] As with any addiction, the developed world's addiction to chemical pesticides will not be easy to break. Just as a drug addict may require greater doses or strengths to achieve the same result, and will often become ill if she or he goes too long without using the drug, our current system of industrialized farming sets up a situation where more and more toxic pesticides are needed to achieve the same pest control results and where failure to apply pesticides can lead to serious pest outbreaks resulting in significant crop losses. This vicious cycle of pesticide use leading to more pest problems and consequently more pesticide use has created a pesticide treadmill. A number of legal, political, and societal changes will be needed to break our pesticide addiction and step off the pesticide treadmill.

While the title of this chapter may be somewhat tongue in cheek, the serious problem of pesticide addiction is comparable to other addictions or chemical dependencies in a variety of respects. Programs developed to break addiction, such as the twelve-step programs created by Alcoholics Anonymous,[3] can provide a useful framework for treating pesticide addiction.

I. STEP 1: ADMIT THAT YOU HAVE A PROBLEM

The first step in any addiction recovery program is to admit that you have a problem. Since the advent of synthetic pesticides starting in the 1940s, and accelerating during the Green Revolution of the mid-twentieth century, the United States has been caught in a vicious cycle of pesticide use, triggering the need for ever greater amounts and types of pesticides needed to achieve the same pest control result. Once synthetic chemical pesticides became widely available, their use skyrocketed. Currently, in the United States alone, approximately 1.1 billion pounds of pesticide products are used annually, comprising 22 percent of the pesticide use worldwide. In the United States, out of the total 1.1 billion pounds of products, more than 850 million pounds are the chemical pesticidal active ingredient in the product.[4] Less than 0.1 percent of pesticides applied in the field actually reach the target pest.[5] The remaining 99-plus percent of the pesticides are released into the environment where they serve no useful purpose and where their residues can travel distances via wind and ground and surface water and ultimately come into contact with humans, wildlife, and other nontarget species.

The risks associated with chemical pesticide use are well established, but perhaps not well understood by the general public. Pesticides are designed to kill living organisms or disrupt natural biological behavior and function, and therefore pose both direct and indirect risks to humans and the environment. Pesticides pose risks to humans through a variety of routes of exposure. The general public is exposed to pesticide residues in food and drinking water. A much greater risk, however, is borne by those who work with pesticides. Farmworkers and their families have direct exposure to significant quantities of pesticides and consequently are at increased risk for pesticide-related health hazards.[6]

Synthetic chemical pesticides not only kill or adversely affect the pest organism they target, but also can have serious impacts on other nontarget species that live in or around agricultural lands. These adverse impacts range from subtle neurological or behavioral impacts to serious reproductive impacts and even death. Some synthetic pesticides adversely affect the reproductive ability of wildlife by mimicking hormones, such as estrogen. Environmental risks from pesticide use are often intensified because of the tendency of certain pesticides to undergo bioaccumulation, a phenomenon in which pesticide concentrations become magnified in the tissues of animals that feed on plants and animals lower in the food chain. Pesticides applied to farm fields can adversely affect wildlife found within the farm boundaries, as well as nearby ecosystems that may be contaminated by pesticide runoff in water, pesticide drift through the air, or movement of contaminated organisms.[7]

Although the general public may be aware of risks of certain banned high-profile pesticides, such as DDT, the risks of many other pesticides currently in widespread use may not be as well understood by the lay public. For example, many organophosphate pesticides, which replaced the banned organochlorine pesticides such as DDT, pose a variety of risks to humans and wildlife.[8] Other pesticides, such as the class of pesticides called neonicotinoids, are linked to deaths of invertebrate species, including economically important pollinator bees and other invertebrates important to agriculture, such as earthworms and other soil-dwelling organisms that create the rich organic soils necessary to produce healthy and hearty crops.[9]

II. STEP 2: ACCEPT WHAT YOU CANNOT CHANGE

Another step in breaking an addiction is to know what you can change and accept what is beyond your control. Decades of attempting to use technology to control nature have taught us that humans will never be completely successful at taming and controlling nature. By way of example, despite dramatic increases in synthetic chemical pesticide use in the United States during the latter half of the twentieth century, crop loss due to insect pests actually almost doubled from approximately 7 percent to 13 percent during that time.[10] There is a certain level of misplaced confidence in our ability to control nature that leads us to believe that we can gain complete control over pests without facing serious ramifications.

This hubris stems in part from the post–World War II obsession with technology, technological solutions, and confidence in human abilities to solve problems through science and technology. However, science and experience make clear that there will always be pests and we will never be able to completely control them. We must recognize that fact and learn to accept some crop loss and damage from pests. An important step in moving away from heavy reliance on chemical pesticides is to acknowledge this fact. Nevertheless, pest management is achievable. There are factors within our control and, while we can never completely eliminate pest problems, we can find ways to manage them effectively.

Today, most industrial agricultural producers engage in prophylactic calendar-based spraying of chemical pesticides as part of their pest management practices.[11] Thus, pesticides are sprayed at scheduled times, whether needed or not. Consequently, large amounts of unneeded and potentially harmful pesticides are released into the environment simply to meet predetermined spraying schedules, without regard to whether the pesticides are actually necessary. More significant, however, is that this overuse of pesticides has led to, and will continue to lead to, more and more pests becoming resistant to pesticides.[12] This is in essence the same phenomenon that is causing the evolution of antibiotic-resistant bacteria, such as MRSA (methicillin-resistant *Staphylococcus aureus*). Antibiotic resistance can result from both overuse and inappropriate use of antibiotics. One common example is when physicians prescribe antibiotics to patients who have illnesses caused by viruses, which are not affected by antibiotics, rather than bacterial infections, which are affected by antibiotics. In both the pesticide and antibiotic scenarios, overuse and inappropriate use leads to the more susceptible individuals in a population of pests or bacteria being killed off, leaving the more resistant individuals unharmed. When these more resistant individuals reproduce, subsequent generations of the pest or bacterial species contain ever greater numbers of resistant individuals and fewer susceptible individuals. Eventually, the population of resistant individuals is so large that the particular pest or bacteria will no longer respond to treatment by that type or dosage of pesticide or antibiotic, creating a treadmill effect where more and more pesticides are needed to keep pace with the increasingly resistant pest population.[13]

The mindless use of pesticides, whether needed or not, has already caused serious pesticide resistance in many species and will continue to do so.[14] With better knowledge of what is actually happening in the field at a particular time, more targeted and less frequent pesticide application may be sufficient. This is the idea behind what is known as ecologically based pest management (EBPM).[15]

One approach to EBPM that has been in use for quite some time is integrated pest management (IPM).[16] IPM is based on the premise that we cannot completely control pests, but we can manage them effectively if we understand the ecology of agricultural systems. IPM has existed for decades, but it is still only used in a limited fashion in modern agriculture. As its name suggests, IPM relies on a combination of pest management tools, while maximizing nonchemical techniques, thereby reducing the problems associated with chemical pesticides. Typically used in IPM is a combination of chemical pesticides, biological controls, and cultural controls, such as crop rotation. A fundamental principle of IPM is that, rather than seeking to "control" pests, it focuses on the more achievable goal of pest "management." IPM was developed in part to combat problems such as development of pest resistance, loss of biological controls, and loss of biodiversity, all created by overreliance on chemical pest control.[17] IPM, in its truest form, demands a thorough understanding of the pest's ecology and a sophisticated scientific knowledge of a site's particular pest problems and available pest management strategies. It also employs the concept of the economic threshold, which describes the point when a site's pest problems reach unacceptable proportions, demanding more intensive pest management strategies, which can include chemical pesticide applications.[18]

An illustrative example of the potential benefits of government policies that encourage IPM occurred in Indonesia in the 1980s. During the Green Revolution, Indonesia had dramatically increased its rice production through intensive chemical pesticide use. However, this intensive pesticide use ultimately led to a major outbreak of the brown leafhopper, which previously had been only a minor pest. The brown leafhopper pest outbreak caused substantial economic losses. Attempts to control the brown leafhopper with more or different chemical pesticides exacerbated the problem. In the mid-1980s, the Indonesian government stepped in and adopted IPM and the reduction of chemical pesticide use as official government policy. Consequently, chemical pesticide use fell by approximately 60 percent. But perhaps more interestingly, rice production grew by approximately 15 percent.[19]

III. STEP 3: LOOK TO A HIGHER POWER

We can learn a great deal about pest management simply by looking at how natural ecosystems function. The phenomenon known as ecological resilience is "a measure of the amount of change or disruption that is required to transform a system from being maintained by one set of mutually reinforcing processes and structures to a different set of processes and structures."[20] One of the most significant components of ecological resilience is biological diversity. A more biologically diverse system is a more stable system in that multiple species serve similar functions.[21] Thus, if one species is lost due to natural or anthropogenic perturbations, other species are available to compensate for the loss of a function previously provided by the lost species.[22] In a naturally occurring ecosystem, a diversity of species, including species that are natural predators and parasites of pest species, tend to keep populations of pest species in check, thereby limiting serious pest outbreaks.[23]

As with natural ecosystems, ecological resilience is also important in agricultural systems. Increasing ecological resilience should be a goal of sustainable agriculture in general, and promoting and maintaining ecological resilience should be an objective of pest management. The presence of a diverse array of species, including natural predators and parasites of pest species, in an agricultural ecosystem provides a complex system of pest control.[24] Modern industrialized agriculture removes much of the species diversity from agricultural systems, thereby eliminating or disrupting natural pest control systems. This can result in population explosions of herbivorous pest species. By attempting to turn farms into sterile industrial facilities free of all pest problems, we have actually exacerbated pest problems. Monoculture production, which consists of large acreages of one crop variety, contributes to pest outbreaks because it provides an unlimited food source for pests that feed on that specific crop variety.[25] For instance, large acreages planted in corn will attract pest species that feed on corn. With a virtually endless supply of food, the populations of these species can grow to levels that produce devastating crop loss.[26] In addition, monocultures, by definition, have low species diversity, and thus, the natural predators and parasites of pest species, which in a natural system would tend to keep the pest populations in check, are absent. The use of synthetic pesticides also eliminates natural predators and parasites of pest species, further disrupting these systems to the point where natural pest control is virtually eradicated in the farm field. In the absence of this natural pest control, pest populations will "flare back," often at higher levels than prior to chemical pesticide application. By retaining and promoting ecological resilience in agricultural systems and creating conditions conducive to enhancing natural pest controls, we can eliminate or reduce the need for chemical pesticide intervention and the adverse human health and environmental effects that accompany it.

IV. STEP 4: EXAMINE PAST ERRORS

There is much we can learn by examining our previous misjudgments with regard to synthetic chemical pesticides. When the pesticide DDT and its relatives came into use in the 1940s, we still believed that we could control nature and use chemicals to gain control over pests. Because these new pesticides worked so well, we believed that they were a panacea that would solve the world's pest problems. Vast quantities of synthetic chemicals were for the first time in history sprayed into the environment throughout the globe. Unfortunately, our optimism was premature, and we soon learned that the pest control benefits of these pesticides came at a great cost. DDT caused egg shell thinning in birds and accumulated in both the environment and the fat tissue of animals, including humans.[27] The early 1970s ban on DDT and the subsequent banning of most of its chemical relative pesticides are credited with bringing many species back from the brink of extinction.

Unfortunately, despite the benefits that have accrued from the cancellation of the registrations for DDT and its organochlorine relatives, the pesticides that we have turned to as alternatives pose serious threats of their own, and many of the same problems persist today with pesticides currently used as replacements for the banned ones. For example, the largest category of pesticides that replaced the banned organochlorines is the organophosphates. While these pesticides do not present the same risk from bioaccumulation that was posed by the organochlorines, they do pose substantial risks to wildlife, including endangered species, due to their relatively high toxicity.[28] They also pose significant risks to farmworkers exposed to them, even when exposure is only through dermal contact. Some newer pesticides, such as the neonicotinoids, have been linked to massive bee killings and are believed to play a role in the worldwide pollinator collapse.[29] As we develop each new type of synthetic chemical pesticide, we continue to believe that it will be better than the previous type and a less risky substitute. While we have eliminated some of the most dangerous pesticides and found some more suitable substitutes, no new chemical pesticide has been a panacea. They all are designed to kill or disrupt living organisms, and all organisms share many of the same physiological attributes that make them vulnerable to many of the same substances. Thus, each new breed of pesticides has its own set of human health and environmental risks. Moreover, any chemical pesticide when overused can result in pest resistance.

V. STEP 5: LEARN TO LIVE A NEW LIFE WITH A NEW CODE OF BEHAVIOR

Weaning our agricultural system off intensive chemical pesticide use will not be an easy task. Society will have to learn to live with a new code of behavior, which will require a multifaceted approach and fundamental shifts in the way we think about agriculture. We will need to find new approaches to agricultural production that continue to produce sufficient food for society and sufficient income for farmers, but are not as heavily reliant on chemical pesticides. Pest management systems that are ecologically based and take advantage of and promote natural pest control systems will have to replace current industrialized approaches that disrupt or eliminate natural pest controls. To achieve this goal, a number of legal and policy changes will be needed, including revising pesticide regulatory laws, rethinking our agricultural subsidy system to redirect subsidies away from industrial commodity production to other crops, and applying more ecologically based pest management practices. More resources will need to be invested in research on EBPM and IPM and education of growers. Moreover, the public will have to change the way it thinks about food, which will require education about the shortcomings of our current agricultural system and a shift away from demanding cosmetically "perfect" food. Consumers will need to support growers who do not rely on intensive pesticide use by purchasing products that may be less attractive in appearance, or in some cases, more expensive. Steps 6–11 outline a number of actions that should be taken to carry out the charge in step 5 so that we can learn to live a new life with a new code of behavior.

VI. STEP 6: CHANGE PESTICIDE LAW TO PROMOTE LOWER RISK PESTICIDES

The primary U.S. statute regulating pesticides is the Federal Insecticide, Fungicide and Rodenticide Act (FIFRA).[30] FIFRA establishes a national pesticide registration system. Generally, pesticides cannot be sold or distributed in the United States without a FIFRA registration. To obtain a FIFRA registration, the registrant, typically the pesticide manufacturer, must demonstrate that, among other things, the pesticide does not pose an unreasonable adverse effect on the environment.[31] FIFRA was never conceived as a true environmental protection statute. Instead it was designed primarily as a national product registration statute, with human health and environmental concerns added almost as an afterthought.[32] Accordingly, from an environmental protection standpoint, FIFRA has a number of shortcomings.[33] A major shortcoming of FIFRA is that it relies on cost-benefit analysis to determine whether pesticides may be sold in the United States. FIFRA's cost-benefit approach is flawed in a number of ways. First, the analysis is conducted on a nationwide basis prior to registration of the pesticide product. Thus, it fails to take into account every specific environmental risk that the pesticide may pose in every particular geographic region. As a practical matter, it is virtually impossible for the Environmental Protection Agency (EPA) to consider environmental risks for every potential place the pesticide could be used, such as the geographic locale of every threatened or endangered species or other ecologically important species, the potential impact on every sensitive ecosystem, or the localized geography and hydrology that could enable environmental long-distance transport of pesticide. By conducting a national cost-benefit analysis prior to registering a pesticide it is possible, if not likely, that many pesticides will be found to have a net benefit on a national basis, while at the same time having the risks outweigh the benefits in particular geographic locales or under specific conditions. It is unrealistic to expect EPA, in the registration process, to consider every possible scenario and conduct cost-benefit analyses for every potential geographic locale and set of circumstances that might exist throughout the country. Another shortcoming of FIFRA is that its primary tool for regulating the use of pesticides is limited to directions on pesticide labels. FIFRA requires a label on each registered pesticide stating that use of the pesticide other than in accordance with the label restrictions constitutes a violation of federal law. However, it is infeasible to provide extensive detailed use instructions and warnings on every label that address every potential risk that may exist for each locale and under every set of circumstances. There is only so much information that can be printed on a product label. When confronted with excessive detailed information, it is likely that many users will not carefully read all of the information or follow all of the detailed instructions completely.

On the cost side of the cost-benefit equation, FIFRA, at least as implemented by EPA, also has limitations in that it does not adequately address the full range of ecological costs associated with pesticide use. FIFRA should be revised to ensure that ecological concerns are adequately calculated in the cost/benefit balancing equation. These concerns include matters such as the degree of uncertainty regarding risks, the level of probability of risk, the extent of the harm that could occur, the reversibility of any harm that does occur, and the likelihood of a pesticide to spread widely in the environment. EPA does not currently fully consider ecological risks at the population level. Instead it focuses its risk analysis on toxicological effects on individuals of the species but does not look at how those effects could impact populations of nontarget species as a whole. Specifically, EPA does not consider the role the affected species population plays in ecosystem function. Further, EPA does not assess indirect impacts to the species, such as how the pesticide affects the species' food sources, predators and parasites, or habitat. Under FIFRA, EPA also is not required to, and does not currently, consider impacts to the ecological resilience of an ecosystem and how it is affected by the addition of the chemical pesticide. A primary way to maintain the ecosystem's resilience is to maintain and promote species diversity. Pesticides that reduce species diversity either in the agricultural system or nearby ecosystems also decrease the resilience of the system. FIFRA should be amended to ensure that ecosystem diversity and resilience is a consideration in any cost-benefit analysis conducted on a particular pesticide.

With regard to the benefits side of the equation, the cost-benefit balancing approach enables EPA to "balance away" significant human health and environmental risks. For example, a pesticide that provides very large benefits on a national basis may be registered even though the risks associated with it are extremely high. Similarly, a pesticide with very low risks may not be registered because the benefits are not high on a national level. Modifications to FIFRA's cost-benefit analysis approach could convert FIFRA into a more environmentally and human health–protective statute. One such modification would be to replace cost-benefit analysis with a standard that would only allow the pesticide to be registered if there were overriding public health, social, or economic benefits. In conducting the overriding benefits analysis on a particular pesticide, EPA would consider all of the potential pesticide or pest management alternatives available. To adequately consider the availability of lower risk alternative pesticides or pest-control methods when making a registration decision, EPA should require applicants to demonstrate that the pesticide is relatively beneficial, either economically or environmentally, over other pesticides or pest-control methods that are available for the particular target pest. The current system does not require a "relative" risk analysis comparing the risks and benefits of each available pesticide or nonchemical pest management method for each pest of concern. Accordingly, under the existing approach, a pesticide manufacturer may obtain a registration for a pesticide even in situations where there are abundant lower risk chemical or nonchemical pest management alternatives available. In other words, pesticides may be registered that pose relatively high risks but have not been demonstrated to have any significant environmental, economic, or societal benefit beyond those already provided by readily available lower risk alternatives. Alternative methods of pest control, including cultural controls such as crop rotation, intercropping, and other approaches, should be considered in determining the true benefits of a particular pesticide.

VII. STEP 7: ABANDON AGRICULTURAL SUBSIDIES THAT PROMOTE INDUSTRIALIZED AGRICULTURE

One policy change that could go a long way to promoting more environmentally friendly farming practices, including practices that do not rely heavily on synthetic chemical pesticides, is to modify the current agricultural subsidy system contained in the U.S. Farm Bill.[34] The Farm Bill contains a number of subsidy programs, many of which provide financial incentives that promote intensive high-yield farming practices. For example, a Farm Bill price support program known as the Marketing Assistance Loan (MAL) program provides nonrecourse loans to help farmers fund their operations. Under this program, farmers receive loans from the government in exchange for placing their crops as collateral. If market prices are high at the time of harvest, farmers can sell their crops and receive the high market price, pay back the loan, and keep the profit. If, on the other hand, market prices are low, farmers can simply keep the loan money and allow the government to take the collateral (the harvested crop) as payment for the loan. In other words, the government is, in essence, paying farmers well above market prices for their crops.[35] Because the loan amount paid is linked to the amount of crop grown, the program incentivizes intensive high-yield production. To achieve such high yields, farmers engage in monoculture farming practices relying on intensive fossil fuel inputs. Because the subsidies are only available for certain specified commodity crops, farmers have no incentive to plant a diversity of crop types or engage in practices such as crop rotation or intercropping, which would help maintain natural pest control, thereby reducing the need for heavy pesticide inputs. This is just one example of the types of perverse economic incentives provided by the current subsidy system in the Farm Bill.

VIII. STEP 8: INCREASE SUBSIDIES FOR ORGANIC PRODUCTION AND EBPM

To move to a more ecologically based pest management system, U.S. Farm Bill subsidies should be shifted away from large-scale industrialized monoculture farming and toward more support for organic production and farming operations that use IPM or other ecologically based pest management practices. Currently the majority of Farm Bill subsidies paid to U.S. growers are paid to large industrialized monoculture commodity crop producers. As described above, many of these subsidy programs reward growers for obtaining the highest per acre yield possible. To obtain high yields, these growers must rely on high fossil fuel input farming practices, including the use of synthetic chemical pesticides. Although there are conservation-based subsidy programs in the Farm Bill, most of those programs, such as the Conservation Reserve Program (CRP) and the Wetland Reserve Program (WRP), involve taking land out of production to preserve it for conservation purposes. Farmers who set aside some land for conservation may feel the need to farm more intensely on the land that remains as working farmland, with fewer environmentally friendly practices and more reliance on chemical pesticides.[36] These programs do not advance ecologically friendly practices including ecologically based pest management on working farm lands.

The two primary working-farm environmental incentives programs in the Farm Bill are the Environmental Quality Incentives Program (EQIP) and the Conservation Stewardship Program (CSP). These programs provide financial and technical assistance to farmers who utilize certain environmental and conservation measures in their agricultural practices. Neither of these programs is intended to reduce pesticide use or encourage the use of lower risk pesticides; however, for CSP in particular, technical and financial assistance is available for certain practices that enhance biodiversity, including protection of natural predators and parasites of pest species. Much could be done to encourage more ecologically based pest control by expanding the CSP program to explicitly provide financial incentives for the adoption of ecologically based pest control methods on farmlands. If subsidies programs were modified to encourage environmentally friendly farming practices, including such practices as EBPM, IPM, and biological or cultural pest control, farmers would likely shift to those practices.

Recent iterations of the Farm Bill have included specific programs designed to encourage and benefit organic producers. For example, the Farm Bill provides funding for organic growers through the National Organic Certification Cost Share program. This program provides payment for up to 75 percent of the annual costs associated with obtaining organic certification from the USDA.[37] Nevertheless, this program provides a maximum annual payment per farmer of $750, far less than many of the commodity subsidy programs that can provide $40,000 or more per year to farmers involved in industrialized commodity crop production.

IX. STEP 9: CHANGE COSMETIC STANDARDS TO REDUCE THE NEED FOR PESTICIDES

Other U.S. government policies that encourage chemical pesticide use include those policies that comprise the cosmetic standards for produce and other foods found in the Food and Drug Administration's (FDA) Defect Action Levels (DALs) and the U.S. Department of Agriculture's (USDA) voluntary grading standards. The FDA's DALs proscribe the maximum allowable level of certain defects in certain specified foods.[38] Many of the DALs have some health-related purpose. For example, some DALs relate to the number of rodent hairs or insect parts allowed in specified volumes of certain foods. However, many DALs are primarily geared toward ensuring the production of cosmetically appealing food. For example, there are DALS that specify maximum allowable levels of "damage" in certain types of produce. The term "damage" in this context includes evidence of pest habitation or feeding. In other words, evidence of insect tunneling or gnawing on produce is considered damage even if the insect is not present in or on the produce. This type of damage, while perhaps not cosmetically appealing, does not in and of itself affect the nutritional value of the produce or the potential for any disease or other health risk to be present in the produce.

The other major government program that encourages pesticide use for cosmetic purposes is the USDA's grading standards program.[39] This program applies to a wide range of fruits, vegetables, and other foods. Although the USDA standards are voluntary, most retailers believe that consumers will only buy produce that meets high grading standards and, therefore, will only purchase produce from farmers meeting these high standards. Many of the criteria necessary to meet USDA high grading standards address purely cosmetic, superficial characteristics, such as discoloration or spots. Although these standards may deal only with superficial cosmetic characteristics, in some cases significant chemical pesticide use is necessary to achieve the standards. The availability of perfect-looking, cosmetically pleasing produce, due in part to the FDA and USDA cosmetic standards, has led Americans to expect their food to look a certain way without discoloration, spots, or other cosmetic blemishes. Thus, although the USDA grading standards are voluntary, they have evolved into what the industry considers to be "required" to meet consumer expectations. A policy shift that no longer requires cosmetic criteria to be met to meet DALs or obtain a high USDA grade would reduce the amount of chemical pesticides used.

Such a shift could have the added benefit of reducing food waste. However, for such a shift to be successful, consumers will need to shift their attitudes about the appearance of the food they purchase. If adequately educated, consumers may be more willing to understand that produce does not have to be perfectly symmetrical, evenly colored, and without any blemish to be healthy, nutritional food.

X. STEP 10: INCREASE FUNDING FOR RESEARCH ON ALTERNATIVE PEST MANAGEMENT SYSTEMS

Promoting the widespread use of EBPM, including IPM, to achieve a more sustainable, resilient, ecologically based agricultural system will require a significant investment of resources. Not only will additional funding be necessary to incentivize these approaches to pest management, but in addition, substantial resources will be needed to fund additional research, education, and dissemination of information needed to bring more environmentally friendly agricultural practices to the farm. EBPM and IPM require a sophisticated understanding of the ecological systems in and around the farm. More research is needed to fully understand the complex relationships between pests and their predators and parasites, as well as their interactions with other organisms and the environment in a variety of ecosystem and agricultural system types. Land-grant universities are logical places to conduct such research and, in fact, already are conducting a fair amount of research in this area. However, in contrast to research conducted on commercial pest control products, such as chemical pesticides that have an obvious economic value, there is little if any obvious economic value to nonchemical pest management techniques. Therefore, additional resources will be necessary for the type of applied research needed to advance EBPM. Once the research is conducted, the information gleaned from the research will have to be disseminated, presumably through agricultural extension service programs. Although many state agricultural extension programs already do include dissemination of information on IPM and other sustainable farming practices, many extension service programs are still heavily involved in their historical focus of disseminating information on chemical pest controls. Because EBPM requires a sophisticated understanding of what is happening in the field at a particular point in time, it will require resources to provide education and training to farmers and farmworkers. While the knowledge and labor needed to carry out EBPM on the farm will require an economic investment, there would be an economic payoff in reducing the financial resources needed to purchase expensive chemical pesticides.

XI. STEP 11: HELP OTHERS

Concerns regarding public health and the environmental risks of pesticides, as well as the sustainability of reliance on synthetic chemical pesticides, are not unique to the United States. In fact, these pesticides often have more serious consequences in other parts of the world that do not have strong environmental laws and where people often have so many critical needs that environmental or long-term health concerns take a back seat to daily survival. Risks from chemical pesticides exist anywhere pesticides are used to combat crop pests or vector-borne diseases exist—in other words, virtually everywhere on earth. Although many of the same issues related to pesticide use in the United States are shared by other countries, not all parts of the world face the exact same challenges. The European Union and other developed parts of the world have industrialized agricultural systems similar to those in the United States. However, in the developing world there are a number of unique challenges. For example, in parts of Africa, West Asia, and Latin America pesticide use is significantly lower than in other parts of the world. Nevertheless, human health and environmental risks in these regions often surpass those in more developed countries. In the developing world, where environmental regulations or enforcement of those regulations may be lacking, farmworkers may have increased exposure to harmful pesticides. People who lack information about the risks of pesticides, or who may not even be able to read labels on pesticide products warning of the dangers, may not use the pesticides properly. One serious concern is that chemical pesticides that are imported into the developing world—sometimes illegally—are often separated from their original packaging and repackaged into containers that do not have labels warning of potential risks or providing safe handling instructions. In fact, in some parts of the world, it is not uncommon for farmers to apply pesticides using milk or water bottles. The use of these seemingly benign containers could cause confusion and potentially cause harm to those who believe a safe food product or water is in the container that actually contains the pesticide. Likely for these reasons, the majority of human deaths from pesticide use occur in developing countries.

Many developing world countries have government policies that, like those of the United States, promote chemical pesticide use by relying on prophylactic calendar spraying to achieve maximum yields.[40] While chemical pesticide use in Asia is widespread,[41] in Africa and Latin America the pesticide use patterns are quite varied. For example, in Africa where most agriculture is done on a small scale, the high cost of chemical pesticides and, in many instances, the limited availability of such pesticides limit their use.[42] Nevertheless, on many commercial cooperatives and larger farms in some parts of Africa, pesticide use is ubiquitous.[43] In Latin America, on the other hand, pesticide use varies greatly from region to region. This variation is due to the substantial geographic, climatic, economic, and cultural diversity throughout Central and South America.[44] In much of the developing world, basic research on pest control and, in particular, research and development on EBPM practices is limited. Even where such research is available, the information is frequently not widely disseminated or used to educate farmers, in part due to organizational challenges and limited financial resources. As one of the wealthiest countries and one of the most technologically advanced countries in the world, the United States is in a position to be a world leader in developing and encouraging EBPM practices that could be shared throughout the world. Research and development conducted by U.S. land-grant universities and other institutions could assist other parts of the world with the information necessary to develop their own EBPM programs.

XII. STEP 12: ENCOURAGE CULTURAL CHANGES THAT PROMOTE REDUCED PESTICIDE USE

Reducing reliance on synthetic chemical pesticides and increasing use of EBPM will require not only the legal and policy changes described above, but also a shift in culture and consumer demand. In recent years the dramatic acceleration of the organic food market,[45] as well as the growing interest in local food and consumers' increased interest in knowing where their food comes from and how it is produced, demonstrate that such a cultural shift is beginning to happen. However, a much more widespread cultural shift will be needed to break our addiction to synthetic chemical pesticides. Consumers will need to cease demanding cosmetically perfect food, and at least in some instances must be willing to pay a premium to purchase food that is grown in a more environmentally sound manner. Providing better education to consumers about where their food comes from and how it is produced can help facilitate this shift. Labeling programs such as the USDA organic food program can help to inform consumers about their purchasing choices. However, food labeling has become overcrowded with labels such as "all natural" that do not have a specific clear meaning. A more standardized food labeling program could clear up some of the confusion associated with the proliferation of labels and could help consumers to make informed choices about the food they purchase and eat.[46]

XIII. CONCLUSION

The twelve-step program set forth above provides a roadmap for stepping off the pesticide treadmill that we have been on for the past fifty years and moving toward a more ecologically based pest management system. A multifaceted approach of combining legal revisions, policy changes, and cultural shifts will be needed to break our chemical pesticide addiction while still allowing sufficient crop production to feed a growing global population. Without these changes, we will continue in the vicious cycle of chemical pesticide use leading to pest resistance, requiring more and stronger chemical pesticides to achieve the same results. Moreover, we will continue to reduce biodiversity in and around our farms, thereby decreasing the natural pest controls that exist and resulting in the need for ever greater pesticide use. As fossil fuels become more expensive and less available, chemical pesticides derived from fossil fuels will also become more expensive and potentially more difficult to obtain. If chemical pesticides were no longer widely available, our large-scale chemical pesticide–dependent monoculture farms would not have the necessary resilience, diversity, and ecosystem function to adapt to nonchemical pesticide pest management. We should be acting now to wean ourselves off synthetic chemical pesticides by developing and encouraging more diversity and more ecological resilience in our farmlands and by improving our understanding of pest ecology and developing better nonchemical pest management approaches. Finally, climate change will likely exacerbate agricultural pest challenges. One particular concern is that a warmer climate will increase pest problems by permitting pests currently found only in the tropics to move into temperate zones. Although climate change will have uncertain impacts on agriculture, there is little doubt that there will be new pest challenges. EBPM can help to combat new pest problems without increased reliance on fossil fuel–derived chemical pesticides.

NOTES

1. Merriam-Webster Dictionary, *available at* http://www.merriam-webster.com /dictionary/addiction.

2. *Biological Components of Substance Abuse and Addiction*, U.S. Congress Office of Technology Assessment 3 (1993), *available at* http://ota.fas.org/reports/9311.pdf.

3. See generally *The Twelve Steps of Alcoholics Anonymous*, *available at* http://www.aa.org /assets/en_US/smf-121_en.pdf. The Alcoholics Anonymous twelve-step approach has been adapted in various forms to a variety of other addiction-related treatment programs. For a description of more than fifty of these programs, see http://www.12step.com/12stepprograms .html.

4. EPA Pesticide Industry Sales and Usage 2006/2007 at 31, *available at* http://www.epa.gov/opp00001/pestsales/.

5. H. F. van Emden and M. W. Service, Pest and Vector Control 41 (2004).

6. Ivette Perfecto and Baldemar Velasquez, *Farmworkers: Among the Least Protected, in* Clifford Rechtshaffen and Eileen Gauna, Environmental Justice: Law, Policy and Regulation 67–68 (2003).

7. For a detailed discussion of the human health and environmental risks associated with chemical pesticides, see Mary Jane Angelo, The Law and Ecology of Pesticides and Pest Management 88–95 (2013).

8. *Id.* at 70–71.

9. See generally Jennifer Hopwood, Scott Hoffman Black, Mace Vaughan, and Eric Lee-Mader, *Beyond the Birds and the Bees: Effects of Neonicotinoid Insecticides on Agriculturally Important Beneficial Invertebrates*, Xerces Society for Invertebrate Conservation (2013); and Jennifer Hopwood, Mace Vaughan, Matthew Shepard, David Biddinger, Eric Madder, Scott Hoffman Black, and Celeste Mazzacano, *Are Neonicotinoids Killing Bees? A Review of Research into the Effects of Neonicotinoid Insecticides on Bees, with Recommendations for Action*, Xerces Society for Invertebrate Conservation (2012).

10. Jason Clay, World Agriculture and the Environment 53 (2004).

11. Helmut F. van Emden and David B. Peakall, Beyond Silent Spring: Integrated Pest Management and Chemical Safety 168 (1996).

12. Van Emden and Service, *supra* note 5, at 115–116.

13. *Id.* at 115–116. For further discussion of pest resistance, see Angelo, *supra* note 7, at 85–88.

14. F. Gould, *The Evolutionary Potential of Crop Pests*, 79 Am. Sci. 496–507 (1991).

15. In 1996, the National Academy of Sciences, National Research Council, published a comprehensive report evaluating and calling for increased use of ecologically based pest management. See National Research Council, *Ecologically Based Pest Management: New Solutions for a New Century* (1996).

16. Van Emden and Peakall, *supra* note 11, at 70.

17. *Id.*

18. *Id.* at 71. For a more detailed discussion of IPM, see Angelo, *supra* note 7, at 77–81.

19. Van Emden and Peakall, *supra* note 11, at 175–176.

20. Garry Peterson, Contagious Disturbance and Ecological Resilience 216, Ph.D. dissertation, University of Florida (1999).

21. *Id.* at 209.

22. Lance H. Gunderson, C. S. Holling, Lowell Pritchard Jr., and Garry Peterson, *Resilience of Large-Scale Resource Systems, in* Resilience and the Behavior of Large-Scale Systems 9 (Lance H. Gunderson and Lowell Pritchard Jr., eds., 2002).

23. Van Emden and Service, *supra* note 5, at 5.

24. Robert F. Norris, Edward P. Caswell-Chen, and Marcos Kogan, Concepts in Integrated Pest Management (2003).

25. Van Emden and Service, *supra* note 5, at 41–42.

26. *Id.* at 41.

27. For a comprehensive discussion of the phenomenon of bioaccumulation of DDT in wildlife, see Bill Freedman, Environmental Ecology: The Impacts of Pollution and Other Stresses on Ecosystem Structure and Function 191–198 (1989).

28. 28   Carl F. Cranor, Legally Poisoned: How the Law Puts Us at Risk from Toxicants 114–120 (2011).

29. Hopwood et al., *supra* note 9.

30. 7 U.S.C. §§ 136–136(y).

31. *Id.* § 136(bb).

32. For a detailed discussion of the history of FIFRA, see Angelo, *supra* note 7, at 113–116.

33. For an in-depth evaluation of FIFRA's limitations, see Angelo, *supra* note 7, at 177–207.

34. The U.S. Farm Bill consists primarily of a set of amendments to the Agricultural Adjustment Act of the 1930s. Congress reauthorizes the Farm Bill every five or so years. The most recent Farm Bill, entitled the Agricultural Act of 2014, was signed into law in February 2014. Pub. L. 113-79. For in-depth descriptions of the Farm Bills' agricultural subsidy programs, see Mary Jane Angelo and Joanna Reilly-Brown, *An Overview of the Modern Farm Bill in* Food, Agriculture and Environmental Law (Mary Jane Angelo, Jason J. Czarnezki and William S. Eubanks II, eds., 2013); and Mary Jane Angelo, *Corn, Carbon and Conservation: Rethinking U.S. Agricultural Policy in a Changing Global Environment,* 17 Geo. Mason L. Rev. 593 (2010).

35. For a detailed summary of the commodity programs in the 2008 Farm Bill, see Jim Monke, *Farm Commodity Programs in the 2008 Farm Bill,* CRS Report for Congress, RL34594, September 30, 2008, http://www.nationalaglawcenter.org/assets/crs/RL34594.pdf.

36. For a more detailed discussion of the conservation-related subsidy programs in the Farm Bill, see Angelo and Reilly-Brown, *supra* note 34, at 14–21.

37. U.S. Department of Agriculture, Natural Resources Conservation Service, *Environmental Quality Incentives Program (EQIP) Support for Organic Growers,* http://www.nrcs.usda .gov/programs/eqip/organic/index.html.

38. U.S. Food and Drug Administration, *Food, Guidance, Compliance, & Regulatory Information, Defect Levels Handbook,* http://www.fda.gov/Food/GuidanceRegulation/Guidance DocumentsRegulatoryInformation/SanitationTransportation/ucm056174.htm.

39. For a discussion of the USDA's grading standards program, see David Pimentel and Hugh Lehman, The Pesticide Question: Environment, Economics, and Ethics 90 (1993). See also http://www.ams.usda.gov/AMSv1.0/ams.fetchTemplateData.do?template=Template A&navID=GradingCertificationandVerification&leftNav=GradingCertificationand Verification&page=GradingCertificationAndVerification&acct=AMSPW.

40. Van Emden and Peakall, *supra* note 11, at 168.

41. *Id.* at 169.

42. *Id.* at 185.

43. *Id.* at 186.

44. *Id.* at 200.

45. See Douglas H. Constance and Jin Young Choi, *Overcoming the Barriers to Organic Adoption in the United States: A Look at Pragmatic Conventional Producers in Texas*, 2 Sustainability 163, 164 (2010) (noting that despite the fact that U.S. organic production has more than doubled over the past two decades, the market for organic products continues to increase such that the United States currently imports products meeting USDA organic standards in order to meet consumer demand).

46. Jason J. Czarnezki, *The Future of Eco-Labeling: Organic, Carbon Footprint, and Environmental Life-Cycle Analysis 4*, 30 Stan. Envtl L.J. (2011); see also Philip H. Howard and Patricia Allen, *Beyond Organic and Fair Trade? An Analysis of Ecolabel Preferences in the United States*, 75 Rural Sociol. 244, 249 (2010).

11

Turning Deficit into Democracy

The Value of Food Policy Audits in Assessing and Transforming Local Food Systems

Caitlin R. Marquis, Healthy Hampshire and
Jill K. Clark, Ohio State University

The prevailing paradigm of food systems change has taken many forms over the years, from the back-to-the-land movement, to a focus on organic standards, and more recently, to the "vote with your fork" movement. One promising paradigm that has taken shape in recent years is the push for food democracy. Food democracy recognizes that, in order to build a policy environment where sustainable food systems can thrive, engaged citizens[1] must know where to start, what to work on next, and how to evaluate their progress. The Food Policy Audit is a tool that can help citizen groups achieve food democracy in their own communities.

This chapter is laid out as follows. First, we provide a general conceptual framework for a local Food Policy Audit. Next, using this conceptual frame, we outline the role of local assessments in evaluating local and regional food systems. Then, we specifically focus on the Food Policy Audit as a citizen-oriented assessment tool, using the food policy council as an example of a citizen-driven group likely to employ such a tool. We use two case studies to detail the value and practical application of the audit tool. Finally, we discuss limitations and opportunities for improving the use of the Food Policy Audit tool to turn local food system deficits into opportunities for greater community engagement in transforming the local food system.

I. THE ALTERNATIVE AGRIFOOD FRAME FOR LOCAL FOOD SYSTEM ASSESSMENTS

The framework for a Food Policy Audit aimed at assessing a local food system is taken generally from the alternative agrifood movement. The label "agrifood" indicates that the movement takes production (agri) and consumption (food) to be inextricably linked. The alternative agrifood framework includes four intertwined themes that aim to counter what is considered the conventional, globalized food system: (1) a systems approach; (2) a sustainability focus (with particular attention to social equity); (3) a place or community orientation; and (4) a democratic and civically engaged process.[2]

A systems approach does not consider production, for example, as separate from consumption. Consumer decisions are said to be "embedded" in ethics and values surrounding food production practices.[3] Conversely, the conventional agricultural approach has been likened to Fordism, where each part of the food system is broken into discrete steps along an assembly line to a finished product, or "commodity," which is undifferentiated from other products of its type. Economic efficiency is the basis for governance of the conventional commodity chain.

The links of the system consider sustainability from not only an economic perspective, but also a social and environmental one. These three central tenets of sustainability are co-considered. This means that sustainability is not a balancing act between economic, social, and ecological considerations, drawing from one, for example, to balance another. Instead it is a multifaceted governance approach that dictates embedding social and ecological considerations in decisions about everything from food production to waste disposal. The agrifood movement has long focused on sustainability as a central tenet.[4] However, friendly critics have rightly pointed out that addressing social equity in the food system is a continual struggle.[5]

A systems approach is often used to understand current problems and identify food system solutions, particularly within a community. For example, Kloppenburg and colleagues used the concept of a "foodshed" (think here, watershed) to identify connections or disconnections in a community-based food system, examining local food sources and community residents within a particular geographic area.[6] In addition, a foodshed could be used as a unit of analysis to envision the potential of a community food system to be self-reliant, drawing boundaries within which to embed economic, social, and ecological relationships between production and consumption.

Concepts of sustainability and systems orientation are practiced and made concrete by place-based community actors. If the globalized food system is both "placeless" and "faceless," then the local agrifood system aims to be otherwise. Many food system initiatives are grounded in local and regional production and consumption, and governed by local decision making. The benefits of food system change that is citizen-driven—rather than market-driven—are to be reaped by community members in specific places and represent the social, ecological, and economic priorities of those community members. For example, community food security (CFS), which is an increasingly common approach to food system work, is a place-based sustainable systems approach that stresses self-reliance and local governance. CFS is defined as "a situation in which all community residents obtain a safe, culturally acceptable, nutritionally adequate diet through a sustainable food system that maximizes community self-reliance and social justice."[7]

The alternative agrifood movement emphasizes progress through democratic participation and citizen engagement. If the globalized food system is privately driven by faceless transnational corporations, then the alternative is a system that is publicly driven by empowered citizens. Instead of consumers, community members are considered "food citizens" by collectively working toward "food democracy."[8] The concept of a food citizen emphasizes individuals involved in civic renewal and engagement, which replace the faceless market that prioritizes economically focused transactions.[9] Hassanein asserts that the alternative agrifood movement is the "main source of pressure" to democratize the food system.[10] The food system agenda is set by food citizens. One example of food system democratization in practice is food system planning, including local food system visioning, planning, and assessment, all of which entail citizen engagement.

Above all, the alternative agrifood movement represents an approach to progress that counters the social, economic, and ecological conditions set forth by the dominant food system. Where the latter is taken by opponents to be faceless, placeless, and holistically unsustainable, the former responds with practices that emphasize economic embeddedness, sustainability, community, and democracy.

II. THE ROLE OF FOOD SYSTEM ASSESSMENTS IN ESTABLISHING FOOD DEMOCRACY

Food system assessments have emerged as a practical response to food system democratization, as place-based evaluations meant to foster communication and collaboration across the food system and provide communities with "next steps." Food system assessments fall under the umbrella of community assessments, which are "activities to systematically collect and disseminate information on selected community characteristics so that community leaders and agencies may devise appropriate strategies to improve their localities."[11] Food system assessments are systems-based and sustainability-oriented, serving to highlight all aspects of a community related to food production, processing, distribution, consumption, and/or food waste. These assessments help "residents, businesses, planners, and food system professionals . . . bring together the diverse interests in the food system to address the complex issues of creating the connectivity and resiliency needed to ensure sustainability across the food system spectrum."[12]

Although literature on food system assessments is scarce, the ultimate goals of assessing a food system typically include (1) documenting community food assets; (2) identifying opportunities for food system change; (3) understanding barriers to social, economic, and ecological sustainability in a food system; and (4) generating comprehensive or targeted food system plans. Achieving these goals relies heavily on citizen involvement, both to identify assets and barriers and to implement resultant recommendations.

A 2011 review of food system assessments in the United States conducted by Freedgood and colleagues analyzed twenty-nine assessments conducted between 1995 and 2011 and divided them into eight categories. Using a food systems frame, the authors focused on assessments that lent themselves to local and regional planning objectives. The emergent categories represented assessments typically conducted by local governments and planning professionals, including the following:[13]

1. Local and regional foodshed assessments
2. Comprehensive food system assessments
3. Community food security assessments
4. Community food asset mapping
5. Food desert assessments
6. Land inventory food assessments
7. Local food economy assessments
8. Food industry assessments

Since food policy audits are a new method of assessment, and the first article on food policy audits was published after the Freedgood review, they are, therefore, not included in the preceding list.[14] Freedgood and colleagues acknowledge that while food system planning ideally encompasses the entire system or life cycle of food, the norm for assessments is to focus on a smaller slice of the system: "hunger advocates tend to focus on food security, public health focuses on obesity, farmland protection groups highlight the land base needed to support local or regional diets, and economists generally concentrate on job creation and economic development."[15] However, the authors also cite a recent trend of comprehensive assessments that aim to create change in all of the five commonly cited food system sectors—production, distribution, processing, consumption, and food waste recovery.

Food system assessments are often carried out to inform a community comprehensive or sector-oriented food system plan. According to Neuner and colleagues, these food system plans "describe communities' goals for their food systems, assess the conditions of food systems, and make recommendations for improving them."[16] Freedgood and colleagues note that community plans emerged from six of the twenty-nine assessments they reviewed.[17] Neuner and colleagues document four additional plans that are based on assessments, evaluations, and community engagement processes.[18] These plans are typically generated or driven by local government entities.[19]

III. THE FOOD POLICY AUDIT AS FOOD SYSTEM ASSESSMENT

While the aforementioned studies serve as valuable context for understanding the role of assessments in food systems change, these studies tend to favor assessments conducted by local governments and planning professionals. However, based on recent data, it appears that a large number of food system assessments are conducted by citizen groups. To understand the level of citizen involvement in food system assessments, it is helpful to look at assessments conducted by food policy councils. According to Harper and colleagues, food policy councils are groups of representatives and stakeholders from many sectors of the food system, whose central aim is "to identify and propose innovative solutions to improve local or state food systems, spurring local economic development and making food systems more environmentally sustainable and socially just."[20] A 2009 study of forty-eight North American food policy councils conducted by Harper and colleagues found that a mere 20 percent of local food policy councils were incorporated into local government, indicating that the majority of food policy councils are citizen-driven. However, a 2012 study of fifty-six councils conducted by Scherb and colleagues found that 94 percent of councils surveyed reported that they had worked to identify problems that could be addressed through policy.[21] These statistics reveal a number of citizen-based groups that have assessed aspects of the food system in their local communities.

Food policy councils often employ assessments to determine where to begin with food system change. A handful of tools, discoverable through a quick Web search for "food system assessment tool," illustrate the instruments available for citizen engagement with local food system assessments. These tools are designed to be utilized by citizen groups, enabling them to evaluate indicators ranging from individual and community food security, to food sovereignty, to food policy. Some examples of these tools include the Michigan Nutrition Environment Assessment Tool,[22] the USDA Community Food Security Assessment Toolkit,[23] the Provincial Health Services Authority's Community Food System Assessment Companion Tool,[24] and the Native Agriculture and Food Systems Initiative's Food Sovereignty Assessment Tool.[25] The Food Policy Audit represents another citizen-driven tool for food system assessment and planning.

IV. ABOUT THE FOOD POLICY AUDIT

The Food Policy Audit (FPA) was developed in 2009 by faculty at the University of Virginia (UVA) and piloted by students in five Virginia counties in 2010.[26] According to O'Brien and Denckla Cobb, the Food Policy Audit was developed to respond to the growing demand for food system assessment tools.[27] The assessment contains 101 points divided into 5 categories—Public Health, Economic Development, Environmental Impacts, Social Equity, and Land Conservation/Access to Land for Food Production—and 20 subcategories. The points are framed as yes/no questions that inquire about the existence of a policy or program in a given jurisdiction. Examples of the 101 points include "Does the locality have an overall wellness plan?" and "Are there land protections for farmers' markets?"[28] The questions were developed largely using food policy framing documents that aggregate and promote best practices in local food system policy and planning. The FPA tool outlines a number of local resources that can be used to complete the audit, including comprehensive plans, zoning ordinances, plans/strategies/programs, stand-alone ordinances, regional or state guidelines, school programming/wellness policies, and school district strategic plans.[29] In a post hoc evaluation of the assessment process, O'Brien and Denckla Cobb also noted the importance of citizen and stakeholder involvement as a local resource utilized during the audit process.[30]

The FPA was adapted for the Franklin County Local Food Council, located in Central Ohio, in 2012.[31] The Franklin County Food Policy Audit (FCFPA) tool was reworked to address concerns specific to the Franklin County Local Food Council. The Food Council was primarily interested in fostering systemic approaches to localizing the food system, preserving agricultural land, and exploring food waste issues. The resultant audit tool contained 100 points divided into 4 broad categories and 18 subcategories. The broad categories included Promoting Local Food, Sustainability, and Community Food Security; Strengthening Zoning and Land Use; Addressing Public Health and Food Access; and Fostering Social Equity.[32] In addition to the questions adopted from the FPA developed at UVA, the FCFPA included unique audit items that were again based on professional documentation of best practices in food system policy and planning.

Table 1 contains a comparison of the categories and subcategories represented by the Food Policy Audit and the Franklin County Food Policy Audit. Both versions of the audit employ the yes/no question and categorical format. One key element that sets the FCFPA apart from the FPA developed by Denckla Cobb and Ray is the employment of a quantified score. With 100 points and a score of 54/100, the Franklin County Local Food Council was able to see that Franklin County was already making a number of policy contributions to the ideal policy climate for thriving local food systems, while documenting opportunities for improvement. In this way, the FCFPA served not only as an agenda-setting tool for local food policy, but also as a quantifiable benchmarking tool for progress toward local food policy goals.[33]

V. SITUATING THE FOOD POLICY AUDIT

While many documented food system assessments have only taken policy change as an implicit or partial goal of food system change, the FPA responds to the need to set specific policy agendas for effecting change throughout the food system.[34] Furthermore, food system assessments are often conducted by professional planners or local governments.[35] These institutions generally have the capacity, knowledge, and resources to design and conduct assessments that draw on extensive research, as well as abundant community and stakeholder input.

The Food Policy Audit, however, more closely reflects an assessment technique adopted by the smart growth movement. Smart growth is a sustainable development framework that emphasizes mixed land use, mixed transportation options, compact building design, housing availability, environmental and open space preservation, walkability, and sense of place.[36] In order to foster smart growth, many localities have developed scorecards to help communities assess successes and opportunities relative to the smart growth principles. The U.S. Environmental Protection Agency notes that smart growth scorecards are

Table 1. Structures of the Food Policy Audit and
Franklin County Food Policy Audit Compared

Food Policy Audit Categories and Subcategories[37]	**Franklin County Food Policy Audit Categories and Subcategories**[38]

1. Public Health
  a. Reduce and prevent community obesity and chronic illness
  b. Engage public by increasing awareness of healthy and local food options
  c. Flexible policies and zoning for creative and adaptive uses
  d. Promote multi-modal transportation options to food sources
  e. Reduce community exposure to pesticides and chemicals in foods
2. Economic Development
  a. Support local food production
  b. Support development of local processing infrastructure
  c. Support development of local distribution infrastructure
  d. Support development of new businesses using locally sourced products & heritage foods
  e. Support increased Security of Food Supply
3. Environmental Impacts
  a. Reduce community carbon footprint and reduce nonpoint source stream pollution
  b. Reduce nonpoint source stream pollution from agriculture
  c. Reduce food waste
  d. Reduce pesticides and herbicides in groundwater and surface water
4. Social Equity
  a. Increase transportation system access to markets that sell fresh and healthful foods by underserved communities
  b. Support location of grocers providing healthy local foods in diverse and underserved locations
  c. Increase availability of fresh and healthful foods for underserved communities
  d. Support an effective emergency food infrastructure
  e. Support equitable working conditions for farm labor
  f. Promote community involvement and ownership in local food system
5. Land Conservation/Access to Land for Food Production
	1. Promoting Local Food, Sustainability, and Community Food Security
	  a. Systemic approaches
	  b. Supporting sustainable agriculture
	  c. Encouraging production for local markets
	  d. Creating markets for local food
	  e. Making local food accessible to low-income populations
	  f. Emergency preparedness and food provisions
	  g. Diverting and recycling food waste
	2. Strengthening Zoning and Land Use
	  a. Urban agriculture on public land
	  b. Urban agriculture on private land
	  c. Home gardening and agricultural use of residential land
	  d. Traditional agriculture and rural land use
	3. Addressing Public Health and Food Access
	  a. Healthy food, wellness, and physical activity
	  b. Food offerings in schools and other public institutions
	  c. Community education and empowerment
	  d. Transportation options for accessing food
	4. Fostering Social Equity
	  a. Food security for disadvantaged populations
	  b. Business incentives for low-income food access
	  c. Equitable conditions for farm laborers

designed to help communities assess their current policies and proposed development projects, and can be completed by citizens with help from planners, consultants, and/or municipal staff.[39] Smart growth scorecards are designed to encourage communities to adopt policies and projects by projecting a readily available set of standards and values against which localities may measure their development. The intent of these scorecards represents a far-reaching and democratized approach to positive change at the community level.

The Food Policy Audit functions similarly in that it establishes a set of goals for communities aspiring to achieve ecologically sound, economically viable, and socially just local food systems. Furthermore, both smart growth scorecards and Food Policy Audit tools function best when reviewed by key community stakeholders prior to conducting the assessments.[40] When carrying out the pre-assessment stakeholder review of the tool, Denckla Cobb and Ray consulted "the community's Obesity Task Force, the regional Planning District Commission, the UVA Health System Nutrition Services, a school system nutritionist, a legal aid advocate for migrant workers, a nonprofit agency serving a low-income neighborhood that was managing the area's first urban farm, and the region's nonprofit agency serving seniors."[41] Likewise, the Franklin County Food Policy Audit was developed with input from the citizen-driven Franklin County Local Food Council, Franklin County Economic Development and Planning, the Franklin County Office of Management and Budget, the Mid-Ohio Regional Planning Commission, and a food policy professor from Ohio State University.[42] In both cases, these stakeholders provided valuable input on relevant content and utility of the audits for the local communities.

As a specific type of community assessment, the Food Policy Audit represents a highly adaptable tool that functions to engage both citizens and stakeholders in an intentional dialogue about local food policy goals, perceived gaps, and potential opportunities. As such, the Food Policy Audit reflects a citizen-based, democratized approach to creating the structural change components of community food security.[43]

VI. THE VALUE OF CONDUCTING A FOOD POLICY AUDIT

Conducting a Food Policy Audit requires familiarity with local regulations, stakeholders, and local food policy goals. Denckla Cobb and Ray fostered this familiarity by piloting the FPA as a student project in a class at the University of Virginia. Students were divided into groups and assigned to audit the city of Charlottesville and the five surrounding counties. The first phase of their project was to review relevant public documents, such as comprehensive plans, strategic plans, school wellness plans, zoning ordinances, regional and state guidelines, and school district strategic plans. Once information was gathered from these documents, students were required to meet with at least five community stakeholders—two from local government—to share their findings and gather feedback as to whether the audit findings reflected reality from the community's perspective.[44]

The Franklin County Food Policy Audit was also conducted using stakeholder input and document review. However, the FCFPA relied more heavily on human capital and stakeholder expertise. At least one contact from the public, private, or nonprofit sector was identified as a potential expert for each audit item. Ultimately, nineteen stakeholders were contacted, and fifteen stakeholders representing thirteen agencies and organizations from the public, private, and nonprofit sectors provided input. The stakeholders interviewed ranged from government officials (Franklin County Economic Development and Planning, Ohio Environmental Protection Agency) to local food educators (Local Matters, Ohio State University Extension), hunger relief agencies (Mid-Ohio Foodbank, Franklin County Emergency Management and Homeland Security), and community and regional development agencies (Economic and Community Development Institute, Mid-Ohio Regional Planning Commission).[45] The process for conducting the audit involved one-on-one stakeholder interviews with a snowball component, wherein each stakeholder was asked for names of others in his or her field who could answer questions posed by the audit. Ancillary public documents, often identified or referenced by stakeholders, were also examined to generate data for the audit.[46]

The benefits of employing both stakeholder interviews and document research in conducting an FPA are many. Stakeholder interviews function to begin dialogues between different sectors around changing the local food policy environment. Often, individuals working in different sectors of the food system are not aware of duplicitous efforts, policies outside of their sector that could influence their work, or potential opportunities for collaboration that exist within the local food system. Additionally, document research allows relevant information from disparate codes, plans, and regulations to be gathered in one place. Since policies and programs that influence the food system are spread throughout a wide variety of departments and sectors, the aggregation of information from each of those sectors represents an important step in identifying food policy gaps and opportunities. Furthermore, the FPA represents a particularly valuable tool for local food policy councils, which are often working to strengthen their relationships to local food-related entities, as well as the relationships between those entities.

VII. PRACTICAL APPLICATIONS OF THE FOOD POLICY AUDIT

In practical terms, the Food Policy Audit strengthens a food system in four key ways: (1) it engages stakeholders from diverse sectors in a dialogue about local food policy; (2) it establishes a vision for an ideal policy environment to support local food system work; (3) it sets benchmarks for future evaluation of the local food policy environment; and (4) it creates a framework around which to set an agenda for improving the local food system. By creating a record of food policy gaps and opportunities in a locality, the FPA serves to support sustainability of food system governance. Whether the results of the audit are utilized by local organizations, local government, local food policy councils, or other entities, the audit creates a comprehensive roadmap for strengthening food policy throughout all of the food system's diverse sectors.

In both the Charlottesville area and Franklin County, the Food Policy Audit has already worked to bring important food policy goals to fruition. In the Charlottesville area, the Food Policy Audit prompted one county to devise methods to assist migrant farmworkers in accessing food assistance programs, and prompted another county to form a Sustainable Food System Council.[47] In Franklin County, recommendations from the audit were prioritized by the Franklin County Local Food Council, and the council identified the passage of "a formal resolution that prioritizes objectives related to public health, ecological sustainability, and economic development with regards to the Franklin County food system" as its first priority.[48] The council then worked with a Franklin County Commissioner and Franklin County Economic Development and Planning to draft Resolution No. 0809-13 "solidifying Franklin County's commitment to a strong and resilient local food system," which passed in October 2013.[49] The resolution, in turn, spurred Franklin County commissioners to solicit a local food economic development plan for Franklin County, as well as to present the audit and associated resolution to other local governments toward strengthening the regional food system. Additionally, the Franklin County Local Food Council held a public listening session to convene stakeholders around its third goal: "Establish a program that increases benefits for EBT [Electronic Benefit Transfer] expenditures at the farmers' market."[50]

While the processes of conducting both the Charlottesville-area FPA and the FCFPA included stakeholder and citizen engagement components, the initiation of the FCFPA was also community-driven. The goal of conducting a food policy audit was identified by the Mid-Ohio Regional Planning Commission's community-driven Central Ohio Local Food Assessment and Plan[51] and adopted by the citizen-driven Franklin County Local Food Council as a primary objective. Additionally, because the Franklin County Food Policy Audit is housed with the Franklin County Local Food Council, the audit receives continued attention and contributes to the sustainability of food system governance. For example, the Policy Working Group of the Food Council uses the audit to develop the working group's annual work plan. The ongoing support of the FCFPA among the Franklin County Local Food Council illustrates the importance of initial buy-in from key community stakeholders in seeing the audit through to implementation of recommendations.

VIII. LIMITATIONS OF THE FOOD POLICY AUDIT

Although the Food Policy Audit offers ample opportunity for food systems change, there are limitations to its use. Some limitations are common to most citizen-driven initiatives, while a few limitations are specific to Food Policy Audits. An example of a common concern is whether decision makers and staff (many of whom are stakeholders interviewed in the audit) buy in to both the process and product. Local officials and staff may or may not be amenable to being "audited." More specific is the concern of where the audit will be housed. The Franklin County Local Food Council serves as a natural home for the Franklin County Food Policy Audit. However, many food policy councils are volunteer efforts and, as with all volunteer efforts, there is always a question of identifying a champion who will lead the use of the audit for policy change. An additional issue involves keeping the audit "alive." The audit is most effective when treated as a living document and should be kept up to date, reflecting changing community objectives and a changing food system. The group that houses the audit may also face limitations regarding its collective knowledge of community needs and objectives. In order to establish audit themes, subcategories, and data collection points, the group will have to identify policy priorities that are appropriate for the particular community setting. Furthermore, when the audit is complete, it will likely reveal dozens of opportunities for policy change. The group conducting the audit will need to have enough local understanding to prioritize these opportunities. Finally, identifying gaps in the food policy environment is not the same as policy analysis or policy evaluation. The group that houses the audit, along with any stakeholders and local officials and staff that they are working with, still has to consider whether a particular policy approach is the right one. An audit can act as a guide to local agenda-setting, but should not be devoid of standard policy practices, such as formulation and legitimation of goals and developing and assessing alternatives.

IX. CONCLUSION

When looking to assess a local food system, there is no shortage of tools and examples for communities to utilize. Our intent in this chapter was not to present the Food Policy Audit as a superior or singular tool to employ in food system assessment, but rather to highlight the role of the Food Policy Audit in achieving food democracy. Given that food democracy promotes citizen decision making in the food system, the Food Policy Audit represents a pathway to claiming that decision-making power by allowing citizens to determine and prioritize the social, ecological, and economic changes they would like to see in their local policy environments.

Using two examples from the Charlottesville, Virginia, and Franklin County, Ohio, areas, we underscored the importance of citizen engagement and comprehensive, systems-based policy examination when determining where to start with changing a local food system. In both cases, conducting the FPA led to examples of food system democratization in the audited communities. While the Charlottesville-area FPA was conducted by students and led to the formation of a citizen group to engage with food system change, the Franklin County Food Policy Audit was conducted by and housed within a local food policy council that was able to use the audit as leverage to carry out policy change in its community. Though evidence is limited, the FPA appears to be an effective tool for achieving a degree of food democracy in a local community.

The two FPAs presented in this chapter are very closely related and simply represent the beginnings of a promising model. Though positive outcomes were seen with the use of these tools, models are never without opportunities for improvement. The Food Policy Audit model certainly has the potential to become as varied and ubiquitous as the smart growth scorecard model. We encourage local communities to build on and strengthen the model by conducting Food Policy Audits in their own localities. We hope that the proliferation of the Food Policy Audit model will begin to take our food systems—both locally and globally—from deficit to democracy.

NOTES

We would like to thank the Franklin County Local Food Council for its crucial role in conducting and fulfilling goals of the Franklin County Food Policy Audit. In particular, we would like to thank Brian Williams, Matt Brown, and Kate Matheny for their initial and continued support of the audit process in Franklin County. We would also like to thank Matt for his commitment to carrying out county-level goals established by the FCFPA. Finally, we would like to thank the following organizations for their valuable input during the FCFPA audit process: Franklin County Economic Development and Planning Department; Franklin County Purchasing Department; Franklin Soil and Water Conservation District; Ohio State University Extension, Franklin County; Economic and Community Development Institute; Franklin County Emergency Management and Homeland Security; Mid-Ohio Foodbank; Local Matters; Ohio Environmental Protection Agency Central District, Franklin County; Ohio Environmental Protection Agency; Mid-Ohio Regional Planning Commission; Franklin County Office of Management and Budget; and Columbus Public Health. Without the support of these individuals and entities, the experience that served as the background for this chapter would not have been possible. This work, in part, was funded by the USDA NIFA Food System Program.

1. While the word "citizen" is used repeatedly throughout this chapter to refer to actively engaged actors in a food system, we would like to acknowledge that food systems rely heavily upon and are relied heavily upon by many individuals who are not "citizens" in the legal sense. We urge food system change-makers to engage these individuals and keep their needs in mind to the greatest extent possible when working toward food democracy.

2. Patricia Allen, Food for the Future: Conditions and Contradictions of Sustainability (1993); Laura B. DeLind, *Are Local Food and the Local Food Movement Taking Us Where We Want to Go? Or Are We Hitching Our Wagons to the Wrong Stars?* 28 Agric. & Hum. Values 273 (2011); Patricia Allen, Margaret FitzSimmons, Michael Goodman, and Keith Warner, *Shifting Plates in the Agrifood Landscape: The Tectonics of Alternative Agrifood Initiatives in California*, 19 J. Rural Stud. 61 (2003); Gail Feenstra, *Local Food Systems and Sustainable Communities*, 12 Am. J. Alt. Agric. 28 (1997); Roberta Sonnino and Terry Marsden, *Beyond the Divide: Rethinking Relationships between Alternative and Conventional Food Networks in Europe*, 6 J. Econ. Geogr. 181 (2006).

3. David Goodman and E. Melanie Dupuis, *Knowing Food and Growing Food: Beyond the Production-Consumption Debate in the Sociology of Agriculture*, 42 Rural Sociol. 5 (2002).

4. Patricia Allen, *Together at the Table: Sustainability and Sustenance in the American Agrifood System* (2004).

5. Allen et al., *supra* note 2.

6. Jack Kloppenburg, John Hendrickson, and G.W. Stevenson, *Coming In to the Foodshed*, 13 Agric. & Hum. Values 33 (1996). http://dx.doi.org/10.1007/BF01538225.

7. Michael W. Hamm and Anne C. Bellows, *Community Food Security and Nutrition Educators*, 35 J. Nutrition Educ. & Behav. 37 (2003) (Emphasis omitted), http://dx.doi.org/10.1016/S1499-4046(06)60325-4.

8. Tim Lang, *Towards a Food Democracy, in* Consuming Passions: Food in the Age of Anxiety (Sian Griffiths and Jennifer Wallace, eds., 1998); Tim Lang, David Barling, and Martin Caraher, Food Policy: Integrating Health, Environment and Society (2009).

9. Patricia Allen, *Reweaving the Food Security Safety Net: Mediating Entitlement and Entrepreneurship*, 16 Agric. & Hum. Values 117 (1999); Feenstra, *supra* note 2; Thomas A. Lyson, *Moving Toward CIVIC Agriculture*, 15 Choices 42 (2000).

10. Neva Hassanein, *Practicing Food Democracy: A Pragmatic Politics of Transformation*, 19 J. Rural Stud. 77 (2003).

11. Kameshwari Pothukuchi, *Community Food Assessment: A First Step in Planning for Community Food Security*, 23 J. Planning Educ. & Research 356 (2004). http://dx.doi.org/10.1177/0739456X04264908.

12. Julia Freedgood, Marisol Pierce-Quiñonez, and Kenneth A. Meter, *Emerging Assessment Tools to Inform Food System Planning*, 2 J. Agric., Food Sys., & Cmty. Dev. 83 (2011).

13. Freedgood et al., *supra* note 12.

14. Jennifer O'Brien and Tanya Denckla Cobb, *The Food Policy Audit: A New Tool for Community Food System Planning*, 2 J. Agric., Food Sys., & Cmty. Dev. (2012).

15. Freedgood et al., *supra* note 12, at 84.

16. Kailee Neuner, Sylvia Kelly, and Samina Raja, *Planning to Eat? Innovative Local Government Plans and Policies to Build Healthy Food Systems in the United States* (2011), 7, http://cccfoodpolicy.org/sites/default/files/resources/planning_to_eat_sunybuffalo.pdf.

17. Freedgood et al., *supra* note 12.

18. Neuner et al., *supra* note 16.

19. Neuner et al., *supra* note 16. Freedgood et al., *supra* note 12.

20. Alethea Harper, Annie Shattuck, Eric Holt-Giménez, Alison Alkon, and Frances Lambrick, Food Policy Councils: Lessons Learned. (2009), 2.

21. Harper et al., *supra* note 20; Allyson Scherb, Anne Palmer, Shannon Frattaroli, and Keshia Pollack, *Exploring Food System Policy: A Survey of Food Policy Councils in the United States*, 2 J. Agric., Food Sys., & Cmty. Dev. (2012).

22. Michigan Healthy Communities Collaborative, *Nutrition Environment Assessment Tool (NEAT)* (2011) *available at* http://www.mihealthtools.org/neat.

23. Barbara Cohen, IQ Solutions, Inc., *Community Food Security Assessment Toolkit*, (USDA Economic Research Service 2002).

24. Christiana Miewald, *Community Food System Assessment: A Companion Tool for the Guide*, (Provincial Health Services Authority 2009).

25. Alicia Bell-Sheeter, *Food Sovereignty Assessment Tool*, (Native Agric. and Food Sys. Initiative, First Nations Development Institute 2004).

26. O'Brien and Denckla Cobb, *supra* note 14.

27. *Id.*

28. Tanya Denckla Cobb and Jessie Ray, *Master Food Policy Audit Template* (2010), http://www.virginia.edu/ien/UVAPlanning_FoodPolicyAudit.pdf.

29. *Id.*

30. O'Brien and Denckla Cobb, *supra* note 14.

31. Caitlin Marquis, *The Franklin County Food Policy Audit: A Report Developed for the Franklin County Local Food Council* (2012), http://www.fclocalfoodcouncil.org/s/FCFPA-Report-Final-w-pics-vasj.pdf.

32. *Id.*

33. *Id.*

34. O'Brien and Denckla Cobb, *supra* note 14.

35. Freedgood et al., *supra* note 12; Pothukuchi, *supra* note 11.

36. Dan Emerine, Christine Shenot, Mary Kay Bailey, Lee Sobel, and Megan Susman, This Is Smart Growth (2006).

37. Denckla Cobb and Ray, *supra* note 28.

38. Marquis, *supra* note 31.

39. U.S. Environmental Protection Agency (EPA), *Smart Growth Scorecards*, Smart Growth (2013), http://www.epa.gov/smartgrowth/smart-growth-scorecards .

40. EPA *supra* note 39; O'Brien and Denckla Cobb, *supra* note 14; Marquis, *supra* note 31.

41. O'Brien and Denckla Cobb, *supra* note 14 at 182.

42. Marquis, *supra* note 31.

43. Pothukuchi, *supra* note 11; Allen, *supra* note 4.

44. O'Brien and Denckla Cobb, *supra* note 14.

45. Marquis, *supra* note 31.

46. *Id.*

47. O'Brien and Denckla Cobb, *supra* note 14.

48. Marquis, *supra* note 31, at 4.

49. Hanna M. Greer, Press Release, Commissioners Approve New Policy Resolution, Solidify Commitment to Local Food System (2013), http://www.franklincountyohio.gov/public/legacy-news/10BCE120-A48D-83C0-FD1E72AD6C06FCE9.pdf; Franklin County Commissioners, Resolution No. 0809-13 (October 22, 2013), http://crms.franklincountyohio.gov/RMSWeb/pdfs/Resolutions/r_000006534/resolution-published.pdf (Resolution Solidifying Franklin County's Commitment to a Strong and Resilient Local Food System).

50. Marquis, *supra* note 31, at 4.

51. Mid-Ohio Regional Planning Commission (MORPC), Central Ohio Local Food Assessment and Plan (Apr. 2010).

Index